The Complete Guide to Alternative Home Building Materials & Methods:

Including Sod, Compressed Earth, Plaster, Straw, Beer Cans, Bottles, Cordwood, and Many Other Low Cost Materials

By Jon Nunan

THE COMPLETE GUIDE TO ALTERNATIVE HOME BUILDING MATERIALS & METHODS: INCLUDING SOD, COMPRESSED EARTH, PLASTER, STRAW, BEER CANS, BOTTLES, CORDWOOD, AND MANY OTHER LOW COST MATERIALS

1405 SW 6th Avenue • Ocala, Florida 34471 • Phone 800-814-1132 • Fax 352-622-1875
Web site: www.atlantic-pub.com • E-mail: sales@atlantic-pub.com
SAN Number: 268-1250

Library of Congress Cataloging-in-Publication Data

Nunan, Jon, 1978-
The complete guide to alternative home building materials & methods : including sod, compressed earth, plaster, straw, beer cans, bottles, cordwood, and many other low cost materials / by Jon Nunan.
p. cm.
Includes bibliographical references and index.
ISBN-13: 978-1-60138-245-0 (alk. paper)
ISBN-10: 1-60138-245-6 (alk. paper)
1. Ecological houses--Materials. 2. Dwellings--Environmental aspects. 3. House construction--Cost control. 4. Green products. 5. Waste products as building materials. 6. House construction--Materials. 7. Appropriate technology. I. Title.
TH4860.N86 2009
690'.8--dc22
2008035306

Printed in the United States

PROJECT MANAGER: Erin Everhart • eeverhart@atlantic-pub.com
INTERIOR DESIGN: Samantha Martin • smartin@atlantic-pub.com
ASSISTANT EDITOR: Angela Pham • apham@atlantic-pub.com
COVER AND JACKET DESIGN: Jackie Miller • sullmill@charter.net

Printed on Recycled Paper

We recently lost our beloved pet "Bear," who was not only our best and dearest friend but also the "Vice President of Sunshine" here at Atlantic Publishing. He did not receive a salary but worked tirelessly 24 hours a day to please his parents. Bear was a rescue dog that turned around and showered myself, my wife, Sherri, his grandparents Jean, Bob, and Nancy, and every person and animal he met (maybe not rabbits) with friendship and love. He made a lot of people smile every day.

We wanted you to know that a portion of the profits of this book will be donated to The Humane Society of the United States. ***–Douglas & Sherri Brown***

The human-animal bond is as old as human history. We cherish our animal companions for their unconditional affection and acceptance. We feel a thrill when we glimpse wild creatures in their natural habitat or in our own backyard.

Unfortunately, the human-animal bond has at times been weakened. Humans have exploited some animal species to the point of extinction.

The Humane Society of the United States makes a difference in the lives of animals here at home and worldwide. The HSUS is dedicated to creating a world where our relationship with animals is guided by compassion. We seek a truly humane society in which animals are respected for their intrinsic value, and where the human-animal bond is strong.

Want to help animals? We have plenty of suggestions. Adopt a pet from a local shelter, join The Humane Society and be a part of our work to help companion animals and wildlife. You will be funding our educational, legislative, investigative and outreach projects in the U.S. and across the globe.

Or perhaps you'd like to make a memorial donation in honor of a pet, friend or relative? You can through our Kindred Spirits program. And if you'd like to contribute in a more structured way, our Planned Giving Office has suggestions about estate planning, annuities, and even gifts of stock that avoid capital gains taxes.

Maybe you have land that you would like to preserve as a lasting habitat for wildlife. Our Wildlife Land Trust can help you. Perhaps the land you want to share is a backyard—that's enough. Our Urban Wildlife Sanctuary Program will show you how to create a habitat for your wild neighbors.

So you see, it's easy to help animals. And The HSUS is here to help.

2100 L Street NW • Washington, DC 20037 • 202-452-1100

www.hsus.org

Dedication

To Mom, Dad, and anyone else who would rather raise a chicken than buy an egg.

Table of Contents

Foreword

By Kelly Hart

The mounting pressure on Earth's ecosystems through human activity is undeniable. In this era of global warming, peak oil, and declining biodiversity, it is clear we must find new ways to live so that future generations will also have accessible opportunities for a healthy life. However, our modern lifestyle is stifling our very chances for survival.

As individuals, one of the areas where we have the most control over energy and resource use is in our homes. Whether building a new house or remodeling an existing one, we can make choices that will diminish our consumption of fossil fuels and lessen the emission of greenhouse gasses, while preserving precious water, forests, and other resources.

The Complete Guide to Alternative Home Building Materials and Methods offers many options for building green and sustainable homes. The alternatives presented tend to reduce the use of energy-intensive industrial materials and promote energy-efficiency in your home over time. The best choice of exactly what to adopt for your own home building project depends on many factors, including local availability of materials, climate, cost, design, ease of construction, and personal aesthetics.

While no book can describe everything you might need to know about every aspect of building, this book does a remarkable job of introducing the basics of alternative home building. It takes you through the step-by-step process of construction, from foundations and roofs to electrical installation and wall systems. Choices for heating, cooling, and insulating your home are also evaluated. The most common – and some uncommon – alternative methods of building are thoroughly examined. What is actually involved in building with cordwood, adobe, light straw-clay, rubber tires, straw bales, earthbags, cob, sod, rammed earth, compressed earth blocks, papercrete, recycled cans and bottles, and salvaged materials? By the time you have finished reading this book, you should have a pretty good idea of how to answer this question.

There are many case studies interspersed throughout the text that amplify the themes presented and provide a personal and pertinent statement about how these ideas are implemented in the real world.

Many resources for further study are outlined to help you down your path of creating a green future for yourself and others. Having gone down this path myself, I can attest to how extremely satisfying it is to actually live in a house that you had a hand in designing and building. And knowing that it is as sustainable as you can make it only adds to the satisfaction. Happy building!

Kelly Hart Biography

Kelly Hart is the founder of **www.greenhomebuilding.com** and has been involved with green building concepts for much of his life. Hart spent many years as a professional remodeler, during which time he became acquainted with many of the pitfalls of conventional construction. He has also worked in various fields of communication media, including still photography, cinematography, animation, video production, and now Web site development. One recent DVD program he produced is "A Sampler of Alternative Homes: Approaching Sustainable Architecture," which explores a range of building concepts that are earth-friendly. Hart is knowledgeable about both simple design concepts and more complex technological aspects of home building that enhance sustainable living. He has even designed and built a solar-electric car that he drives around his neighborhood. Hart and his wife, Rosana, built an earthbag/papercrete home in the mountains of Colorado.

www.greenhomebuilding.com

Preface

If it were not for my mother's lifelong obsession with natural living and my father's uncanny ability to underestimate the difficulty of just about everything, I probably would not be writing this book. In the fall of 1995, my mother began casually dropping hints that she would like to build an eco-friendly cordwood house with her own hands; sometime in the early part of 1996, my father decided that this was a fine idea and began making preparations to sell our current home and begin construction of the house they live in today.

At that point, none of us had any idea what we were doing. If my father had ever picked up a hammer to do anything more than hang a picture, I was certainly not aware of it. The good sense and intricate planning that generally go into the construction of a family's primary residence were decidedly absent from the very beginning; the "blueprint" my father sketched on college-ruled notebook paper with a blue ballpoint pen was – and remains – the punch line to many of our friends' and family's most-amiable inside jokes. Despite the lack of any formal knowledge of construction and with only our six hands to do the heavy work, Mom, Dad, and I (with some help from my 11-year-old sister and 6-year-old brother) began building our new home in June 1996.

Things did not go well at first. The plan was to live in tents all summer and have the house ready by the time Pennsylvania started to get too chilly to sleep outside, which is generally sometime around mid-September. The tents blew over the first week, so Mom, Dad, and the kids moved in with my grandmother in the next town over, and I moved myself into my father's hatchback Suzuki® Swift – which, at 18, felt like my own alternative dwelling. The 10-hour days we put in (when not encumbered by day jobs or other obligations) certainly bore fruit, but by early August, it became very plain that we had seriously underestimated how long this sucker was going to take to construct.

To make a very long story short, the house was habitable after about six months, although it was far from fully finished. Our experience was not unlike the experiences of many owner-builders, and even though the job did not go exactly as planned, the finished product is something each of us is extremely proud of. The obstacles that popped up when building the house my parents now live in were numerous, and so were the mistakes we made along the way, but perseverance and patience carried us through. Now, we have a heck of a house. Like many owner-builders, we skimped on the planning and ended up paying for it both financially and in terms of effort; in the end, however, everyone involved in the construction of our cordwood home is glad we took the alternative building plunge.

No one ever said it was going to be easy – except for Dad, of course. It is my sincere hope that this book will help anyone who reads it to understand the kind of undertaking building an alternative home really is and simultaneously illustrate what a fulfilling and unique process it can be when done right (or even when done halfway right). Anyone who is interested in this kind of construction already has the majority of the tools necessary to bring their dreams to life in the form of their own bodies and minds. However – and trust me on this one – understanding the other tools and techniques required to build a sustainable structure before you move out of your current residence to start building is not a bad idea, either.

Introduction

It would be misleading to say that one book can explain everything there is to know about alternative building. The world of alternative, sustainable, and green building is enormous, and a book that could tell you everything would be next to impossible to compile — and rather cumbersome to carry around. Many of the books written about alternative building focus on one or maybe even a few important facets of creating a house from scratch; while these books are certainly useful, it would take quite a few of them to really get a good understanding of the entire process. This book attempts to remedy that.

Unfortunately, one book cannot tell you every option for every step in every situation; what one book can do, however, is give you a few options for the likely steps you will find in most situations. This book will describe these steps. It will give you choices at every step, describe the benefits and drawbacks of these choices, and give you suggestions as to which choices might make the most sense in your particular situation. It is my sincere hope that anyone who reads this book will know the benefits and drawbacks of building an alternative structure, understand the basic steps necessary to build such a structure, and have a good idea of which type — if any — of these building techniques will best suit his or her situation.

If you are interested in becoming an owner-builder, this book will help you better plan your alternatively constructed home. You will find examples of what materials and techniques work best in your area, as well as advice on how best to procure these materials as cheaply as possible. Remember, though, that building a house is a huge project that cannot be completed overnight. If you are looking for a cheap place to live and need it tomorrow, you will have far better luck looking in your local classifieds. Also remember that no structure that is to be equipped with modern conveniences is ever built by a single, first-time builder. If you want the kind of home you would find in a housing development, you will have to buy one; if you simply want a hole in the ground, you hardly need a book to help you shovel. If you want a realistic description of how an owner-built, alternative dwelling is created, however, you have come to the right place.

What Is Alternative Building?

In many cases, the types of building that are considered alternative today have their roots in structures that humans have been constructing for centuries. The reason why these building techniques are gaining popularity in modern times is twofold. First, these old building techniques are far more eco-friendly than the majority of structures we are used to seeing; second, these structures are simple enough in nature that they can be built cheaply and without the aid of a lot of the heavy and expensive equipment associated with most new construction.

Some alternative building techniques are far from ancient. Constructing a home from used tires, for example, is a process that could have only taken place after the invention of rubber tires in the mid 1800s. In the end, however, it does not matter too much whether the technique is old or new; what really matters is that these techniques exist and are being used today by people like you.

The terms green, sustainable, and alternative get thrown around a lot these days, and it can get quite confusing in some situations to tell whether one particular technique falls under one or more of these headings. In this book, green building will refer to any construction practice that is less damaging to the environment than a similar practice used in conventional lumber-framed construction. Sustainable practices will encompass any building technique that can be done repeatedly without changing the environment in any noticeable way. Alternative building will denote any practice that you do not see in conventional construction.

For example, creating a structure from matchsticks would be considered alternative building; if the energy used to power the electricity in the matchstick structure came from solar panels, the structure could be considered both green and alternative; if the matchsticks used in creating the structure were produced by a company that replants the trees that they harvest or were pulled from a landfill rather than purchased new, the alternative, green structure that you end up with would also be sustainable.

As the owner-builder will quickly find when constructing his or her dwelling, one of the biggest challenges in alternative building is the noticeable lack of options offered among conventional materials and techniques. The construction of your average conventional house certainly shares plenty with that of its alternative cousins, but because the market is so saturated with general guidelines meant for conventional builders, those interested in alternative building have to sift through a lot of unusable advice.

CASE STUDY: ALTERNATIVE CHOICES IN A CONVENTIONAL DISCIPLINE.

Christopher Barton, Founding Director
Sage Mountain Center
www.sagemountain.org
smc@sagemountain.org

CLASSIFIED CASE STUDIES™
directly from the experts

Christopher Barton, founding director of the Sage Mountain Center, has dedicated his time and effort to helping those in search of alternatives to conventional building materials and techniques.

"As an education center, our focus is to promote the sustainable building principles and techniques," Barton said. "Through our workshops, tours, and consultation services, we show others what is possible and how to do it. We have done all of the research and construction of our facility ourselves and continue to explore new ways to advance the technologies. Sage Mountain Center is a harmonization of old-tech and hi-tech."

Now, you do not have to subscribe to vegetarianism or Buddhism to build a house; however, he believes that creating an institution that offers anyone interested in a variety of options for better living is a great way to reach out to the community on multiple levels. Often, those interested in an alternative to one aspect of living are also curious about different ways to approach other facets of their lives.

CASE STUDY: ALTERNATIVE CHOICES IN A CONVENTIONAL DISCIPLINE.

"Sage Mountain Center is an education and demonstration facility dedicated to promoting inner growth, physical health, and sustainable living through education, experimentation, and creative innovation," Barton said. "Our courses have included Straw Bale and Cordwood Construction, Solar Electricity, Wind Generation, Vegetarian Cuisine, Hatha Yoga, Edible and Medicinal Herbs, Meditation, and more."

He understands that teaching by example is a key element to having a successful learning environment. This is one of the reasons why the center itself was constructed using alternative materials and sustainable techniques.

"Sage Mountain Center (SMC) is a cordwood and straw bale building designed to passively garner the light and heat of the sun," Barton said. "With R-values of 28-48 [very high insulation capabilities], the building stays cool in summer so that no cooling methods are needed other than closing south-facing blinds for a couple weeks in the fall. It is off-grid and powered by solar energy. SMC uses one-fifth of the energy normally used for this size of a building, and that one-fifth comes entirely from a hybrid solar and wind electric system. Our solar array and wind generator produce on average 14 kWh of power per day. SMC uses compact fluorescent lighting (CFL), light emitting diode (LED) lighting, and solar light tubes, while the southern orientation of the house provides passive lighting and heating.

"There are four pre-heating water systems at SMC implemented to reduce the use of propane," Barton continued. "First, hot water lines run through a gray water [non-toxic, but not drinking-quality water] warming tank. This preheats water. When we build a fire, hot water lines in the firebox heat water. When our batteries from our solar electric system are full, excess energy is sent to a DC electric water-heating element. And finally, our solar thermal panels on the roof provide 60 percent of our hot water year-around. Our guesthouse, which also has these systems, is completely propane-free. Compared in square footage to a conventional house, our main education center uses about $10 of propane per month. This propane is for space heating via an in-floor heating system, for water heating via an instantaneous water heater, and for cooking on our gas stove. A solar oven is also in use to offset propane used for cooking."

If you asked Barton why alternative building is worth the kind of attention and dedication he has obviously put into learning about it, he would probably tell you that it simply makes better sense than conventional construction.

CASE STUDY: ALTERNATIVE CHOICES IN A CONVENTIONAL DISCIPLINE.

In fact, I am sure that it is his sincere hope that someday, alternative building will be the norm rather than the exception.

"The choices that we make by buying organic food, driving a [Toyota] Prius, and supporting environmental organizations are all part of the alternative lifestyle," Barton said. "We really don't like the term 'alternative' because it connotes something unproven, weird, strange — something to cause suspicion. The fact is that alternative buildings — in the way that we are addressing them — are very well-proven and very sound in their approach and technology."

The Importance of Green and Sustainable Building

As the world population continues to grow at an alarming rate, people are realizing that the planet cannot sustain such continuous and exponential growth. Space is increasingly limited, and we are continually diminishing our natural resources such as timber – much of which is cut to build homes. It is obvious that we cannot continue to build new homes at this rate indefinitely. Awareness of this extreme imbalance has created a trend toward more sustainable living practices, as well as a trend toward building smaller and more sustainable homes. From the ground up, alternative home building is much more environmentally sustainable than conventional home building. Depending on the type and amount of sustainable materials used, green homes can significantly decrease the footprint we humans are leaving on the environment.

Typical modern buildings consume an enormous supply of natural materials. For example, according to the World Watch Institute, they use 40 percent of the world's energy and 55 percent of all the wood cut for non-fuel purposes. About half of the energy for constructing and operating buildings serves heating, cooling, lighting, and ventilation; in the U.S., more energy goes to building and operating buildings than to transportation or to industry. However, green and sustainable construction takes these factors into account from the very beginning; in many cases, an alternatively constructed, owner-built home will not only reduce waste during the building process, but will reduce energy consumption over the home's lifetime.

In terms of home building, conventional construction produces a significant amount of waste; the typical new construction project averages 3.9 pounds of waste per square foot. It is estimated that more houses will be built in the next 50 years than have been built throughout all of human history, which is a lot of potential waste.

Materials used for home building in the last 100 years are typically designed to last for a much shorter period of time than many sustainable building materials, thus adding to the burden of already-full construction and demolition landfills. The waste in these landfills, called C&D — or construction and demolition — is the largest type of waste in the entire human waste stream. Buildings are demolished on a regular basis to clear the way for a bigger store, house, or office building. Demolition creates an enormous amount of waste, averaging 155 pounds per square foot.

In the past few decades, more attention has been focused on the effects of conventional building practices on natural resources. Today many organizations, communities, and individuals see sustainable home building as a means to decrease our ecological footprint. Green building has become a viable solution both to individual home builders and to buyers. Additionally, using local, renewable, and recycled materials encourages other sustainable practices such as conservative energy use, thus focusing on both the present and the future.

Using Readily Accessible Materials

The most important thing to remember when choosing what building technique is right for your situation is this: It is both cost-effective and feasible to work with what is readily available in your area. Building with logs might be aesthetically attractive, but if you live in an arid area, it will cost you unnecessary time and energy when compared to building with materials that are locally abundant. Additionally, it is almost always the case that using easily accessible materials is also a more eco-friendly option than having materials shipped in.

Because the term "readily available" can mean drastically different things depending on where you plan to build, it is important to be flexible when planning your project. If your goal is to construct an eco-friendly dwelling, it is a bit counter-productive to have all of your materials shipped in. Transporting materials from far away requires a great deal of energy consumption and is often far more costly than using what is already present in your area. While a house built out of red pine might be an eco-friendly option in upstate New York, it would be a huge hassle and unnecessary expense in an area like lower New Mexico. Having your heart set on one material or another can become both financially and environmentally costly; when planning your project, it is a better idea to draw inspiration from the materials you see around your neck of the woods rather than from magazines, books, and Web sites.

If you are thinking about building an alternative home, it is a good idea to understand not only the techniques that will be involved during construction, but also the different forms of building that fit well with your particular environment. This book will teach you both.

Chapter 1

Planning and Permits

While not every locale in the United States operates according to building codes — some locales outside city limits have absolutely none — a large portion of the design and building industry operates according to the International Building Code (IBC) that the International Code Council (ICC) developed in the 1990s. Though some builders might still refer to the Uniform Building Code, which influenced the IBC, it has since been rendered defunct. The stated goal is the protection of the health, safety, and welfare of people through their buildings and communities; however, the code is fairly limited to certain types of homes and does not touch on many of the building practices explained throughout this book.

The focus of the IBC is primarily wood-frame homes. Modern building codes often certify alternative materials used as infill, such as straw bale, but often ignore the use of these same materials when intended for load-bearing and structural uses. Such a lack of regulation for sustainable housing should not be a deterrent, however. An engineer can help prove the design's structural integrity, and some builders cite tests for corroborating structural safety. Many go so far as to provide an extensive narrative, clarifying the design and any questions the inspector may have. After the home is built, certain organizations can certify the home's sustainability, which is yet another means for connecting with other green builders, and it is useful when it comes to ensuring many of the qualities a home builder hopes for from energy-efficiency to a healthier home.

As far as getting a building permit to begin constructing a green home, it really depends on where you live. Some areas do not require a permit, and codes vary. Sometimes the size of a building can make all the difference between needing and not needing a permit. For example, in some rural areas, small buildings might not require a permit at all, while the design for a sizable building might be difficult to get approved. The inspector of the local jurisdiction, which varies with locale, approves the design plans, and inspectors typically follow the IBC. If there is any discomfort with the design of a sustainable home, some home builders succeed by providing a legal letter stating that the jurisdiction, which grants the permit, is relieved of responsibility for the methods used to build the home. Be sure to have as much documentation as possible for the inspectors, who ultimately give approval.

Where do you go to get documentation to show the feasibility of alternative building practices? Different organizations, government-operated and otherwise, exist to certify a building's green qualities. Such certification can further help the home builder educate him or herself about different energy-efficient technologies. Though there is no single, federal entity to certify a building's green qualities, the Leadership in Energy and Environmental Design (LEED) tends to oversee individuals, organizations, and businesses for certification. Other groups, such as the Architecture 2030 Initiative and the Green Building Initiative also certify a building's greenness. Having your home's design certified by any or all of these groups may go a long way in helping local authorities and put you on the fast track to design approval.

The LEED Web site (**www.usgbc.org/leed**) offers information on their rating systems, project certification, professional accreditation for builders who wish to certify buildings, and various other resources. The organization is categorized as a nonprofit organization in the U.S. Green Building Council, an umbrella organization for more than 15,000 building organizations, including real estate developers, architects, designers, and contractors. They rate such aspects as design methods, a home's placement in relation to the larger community, sustainability, water and energy efficiency, and indoor air quality. In the first half of 2008, 10,250 homes had registered for LEED status, which was up more than 7,000 since the program began in 2006.

Platinum is the highest ranking for a LEED-certified home, as it earns up to 69 points in its certification process. The lowest LEED ranking that is still certifiable tops out at 32. The lengthy application process for attaining a level of LEED certification requires an on-site inspection and may include paying a fee, depending on the building. It is no small investment in time or energy, but such a process might prove worthwhile for the owner-builder interested in attaining certification.

Architecture 2030 (**www.architecture2030.com**) is a U.S. non-profit, independent organization whose mission is to convert the worldwide building sector from a main supplier of green-house gas emissions to a part of this problem's solution. The organization focuses on getting the building industry, as well as the nation, to adopt their 2030 Challenge: the global initiative for all new buildings to drastically reduce their GHG emissions to 50 percent through fossil-fuel consumption by 2010 and make new buildings carbon neutral by the year 2030. The U.S. Green Building Council is one of the organizations that has adopted the 2030 platform.

The Architecture 2030 Initiative offers a prescription for reducing home energy consumption, which includes passive solar design, high insulation, tight air sealing, and using daylight more effectively. It strongly advocates using alternative energy technologies and purchasing renewable energy, along with using efficient products.

Dealing with Building Inspectors

Depending on where you plan to build, your construction may need periodic approval by a building inspector. If your area has specific building codes for your style of construction, like California's codes for straw bale walls or the many areas where adobe brick strength will have to be tested, a building inspector will need to be called in to approve certain aspects of your project. If you are building in or near an urban area, you can almost guarantee that a building inspector will play a role in your project, as urban areas have more (and more stringent) rules about how a building can be constructed. Because laws vary dramatically, it is important to find out beforehand at what stages a building inspector will need to be called. Failure to do this could result in some unnecessary backtracking, which will cost you money, time, and effort that could better be used elsewhere. Failure to follow code can also result in serious problems when you are looking to insure your property.

There is a chance that your building inspector will have never dealt with the type of structure you plan on constructing. For some, inspecting an alternative building will be a novelty; for others, it might be intimidating. A building inspector working on an alternative home will be able to identify all of the familiar elements of the structure, but because alternative building is very different from what most inspectors deal with on a daily basis, it can be overwhelming. Remember, building inspectors are not there to hassle you; they might have a lot of questions about your structure and may require you to conform to certain rules, but, ultimately, their job is to make sure that the structure you build is a safe one. While some codes and practices might seem like a pain, restrictions are in place for a reason. By creating a congenial relationship with your local building inspector, you will make his or her job more enjoyable, and you might even get some helpful advice in the process.

Having an experienced home builder on your team or available for consultation can help you keep track of the codes in your area and when it is appropriate to call in the inspector. One of the nice things about constructing an alternative home is that other alternative builders are often willing—even excited – about helping first-time builders. Today, the Internet makes it even easier to find other alternative builders to talk to. It is definitely worth the time to look for others who have been through this process before; cob, straw, and cordwood have built as many friendships as they have homes.

Seeking Professional Help

When you first begin looking at pictures of alternatively constructed homes and owner-built houses, you are bound to notice some pretty striking differences between these structures and the professionally constructed homes you are probably most familiar with. In the green and alternative building world, it is not uncommon to see brand new homes fitted with windows and doors that look like they were pulled from a landfill. Anyone considering constructing this style of home should understand up-front that prim and polish is not the main goal of most owner builders. Unlike a stick-frame house in a new housing development, green and alternative owner-built structures are as different from one another as the people who construct them; few houses are structures that will look as flawless and airbrushed as the "cookie-cutter" houses you will find in new housing developments.

Despite – and because of – the numerous eccentricities you will find in an owner-built home, most who take the plunge find that they cannot help but love the house they live in. Nearly all will attest that the process of creating their own permanent dwelling with their own bare hands was a delightful (though admittedly strenuous) one. There are some folks out there who will look at a usable and soundly constructed owner-built structure and see only the differences between the building in front of them and their idea of what a house should be. The folks who reside in these left-of-center houses, on the other hand, are aware that a house does not have to look a certain way to be a fine place to live.

The fact is that many of the people who opt for alternative building techniques have little or no prior building experience. While there is certainly something to be said for building techniques that are simple enough for a first-time builder to employ successfully without any outside help, it is also true that many of these structures end up having plenty of character. Most people who decide to build their own house out of logs, straw, or mud are already aware that the finished product will be a little off-center; if this is not your idea of a dream home, you should probably start checking your yellow pages for general contractors.

If you want professional results in both appearance and function, you may have to make some compromises. In most cases, that compromise will take the form of researching for hours, learning by doing, and getting through trial and error. Unless you are already well-versed in every step of the building process, you are going to make some mistakes. This is, of course, normal. When your errors are simply aesthetic, there is a good chance that when your home is ready to live in, you will have plenty of chances to go back and spruce up your earliest, least-experienced handiwork. However, when your mistakes are more than skin-deep, they can have a more dramatic effect on your home's function and longevity.

Many owner-builders have found that the quickest way to a successful project is often a combination of their own efforts and a sprinkling of professional attention; in many sections of this book, it is recommended that you seek professional help for some stages of your building project. The best scenario for most owner-builders would be having a friend or family member who is experienced in construction give free advice and consultation. But this may not be the norm.

Between the two opposite ends of the construction spectrum — a completely owner-built house and a completely professionally built house — is where most owner-builders will find themselves. There will be things that you are simply not able to do on your own, but there will also be many that you would be crazy to hire a professional crew to do. Most alternative building techniques are labor-intensive, and hiring a professional crew for simplistic tasks, like to mix cob with their feet or apply earth plaster to the exterior of a straw bale wall, will get expensive quickly. On the other hand, hiring a professional to help you with the fundamental tasks like laying a foundation, installing electrical wiring, creating a structurally sound design, or putting in a septic system is often the best, safest, and fastest way to get quality results.

Getting professional results without a professional on your team can be accomplished, but it is important to understand that no amount of research or planning can take the place of years of on-the-job experience of a professional (or at least well-seasoned) builder. If you plan to go it completely on your own, this book will give you the information you need to create a basic dwelling, but unless you have a great deal of knowledge about construction, do not expect to build the Taj Mahal all by your lonesome. Without the help of some professionals, owner-built homes are often small, simple structures, and many of them lack certain systems that modern homeowners view as necessities. Larger owner-built homes that contain all the comforts of any other modern structure certainly exist, but they are typically built by folks who had previously constructed a modest dwelling or two.

Many of the methods described will vary greatly from one situation to another, so be prepared for trial and error. Professional construction companies are not likely to be well-versed in many of these techniques, either, but a professional will still be aware of how the conventional components of your house should work. There is certainly no shame in requesting the advice of a professional. Even if you do not have a friend or relative in the construction business, consultations with professional builders along the way are unlikely to be significantly costly – particularly if you need advice rather than labor. Ultimately, the choice to seek professional counsel is yours, but if you are looking for the most effective path to a functional house, it is recommended.

CASE STUDY: OWNER-BUILT, PROFESSIONALLY DESIGNED.

Sigi Koko, Owner and Principal
Down to Earth Design
www.buildingnaturally.com
sigikoko@buildnaturally.com

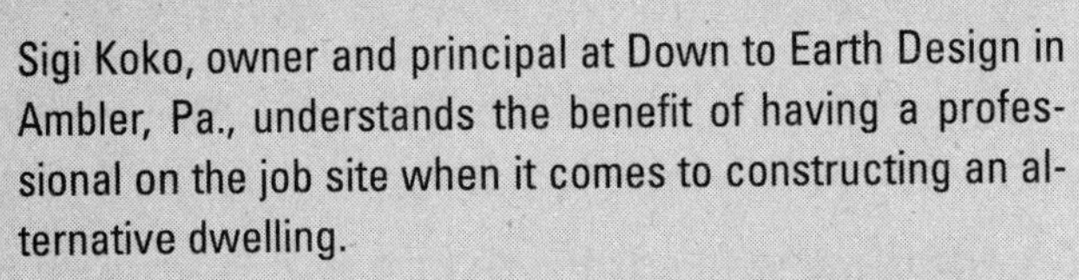

Sigi Koko, owner and principal at Down to Earth Design in Ambler, Pa., understands the benefit of having a professional on the job site when it comes to constructing an alternative dwelling.

"I design straw bale homes and teach workshops on many natural building techniques," Koko said. "I founded the company in 1998 because I wanted to design straw bale homes and couldn't find an architecture firm that did so at the time."

Owner-builders looking to build a straw bale structure have many challenges to face, but one of the least well-known is this: Though straw and other alternative material might seem inexpensive, many who build with them often end up paying just as much for their homes as anyone else due to poor planning. Down to Earth Design works with owner-builders to make sure this does not happen.

"I design all of my buildings on the same or lower budget as 'conventional' homes, and the long-term energy performance means that operating costs are reduced by an average of 50 to 80 percent for homes I design," she said. "These homes far outperform 'conventional' buildings in terms of indoor air quality, energy efficiency, and quality of construction. Additionally, they provide the unique opportunity for owner participation during construction, which I believe connects them more closely with their home."

Koko has heard her share of skepticism when it comes to what she does, but is confident that she and other alternative building firms are on the right track.

"I have seen more knowledge in the general public, more knowledge on the part of permitting officials, fewer 'Three Little Pig' jokes — there is generally less skepticism and more embracing of 'alternative' techniques," she said. "What used to be 'alternative' is becoming mainstream."

There are certainly challenges to building with straw in Pennsylvania's (and in general, the whole Northeast's) demanding climate, but Down to Earth is making sure that its homes will perform better than any conventional home in the area.

CASE STUDY: OWNER-BUILT, PROFESSIONALLY DESIGNED.

"Though there seems to be the perception that straw bale only works where the climate is arid, it is feasible to build durably with straw bales in wet, humid climates," Koko said. "However, alternative homes can be more time-consuming to build (more labor intensive), and they usually require educating permitting officials to show how 'alternative' construction meets the existing building codes.

In short, owner-builders can certainly be successful on their own, but the most successful projects often have one thing in common: "someone on the team during the design process who has experience with the alternative technique(s) being used," Koko said.

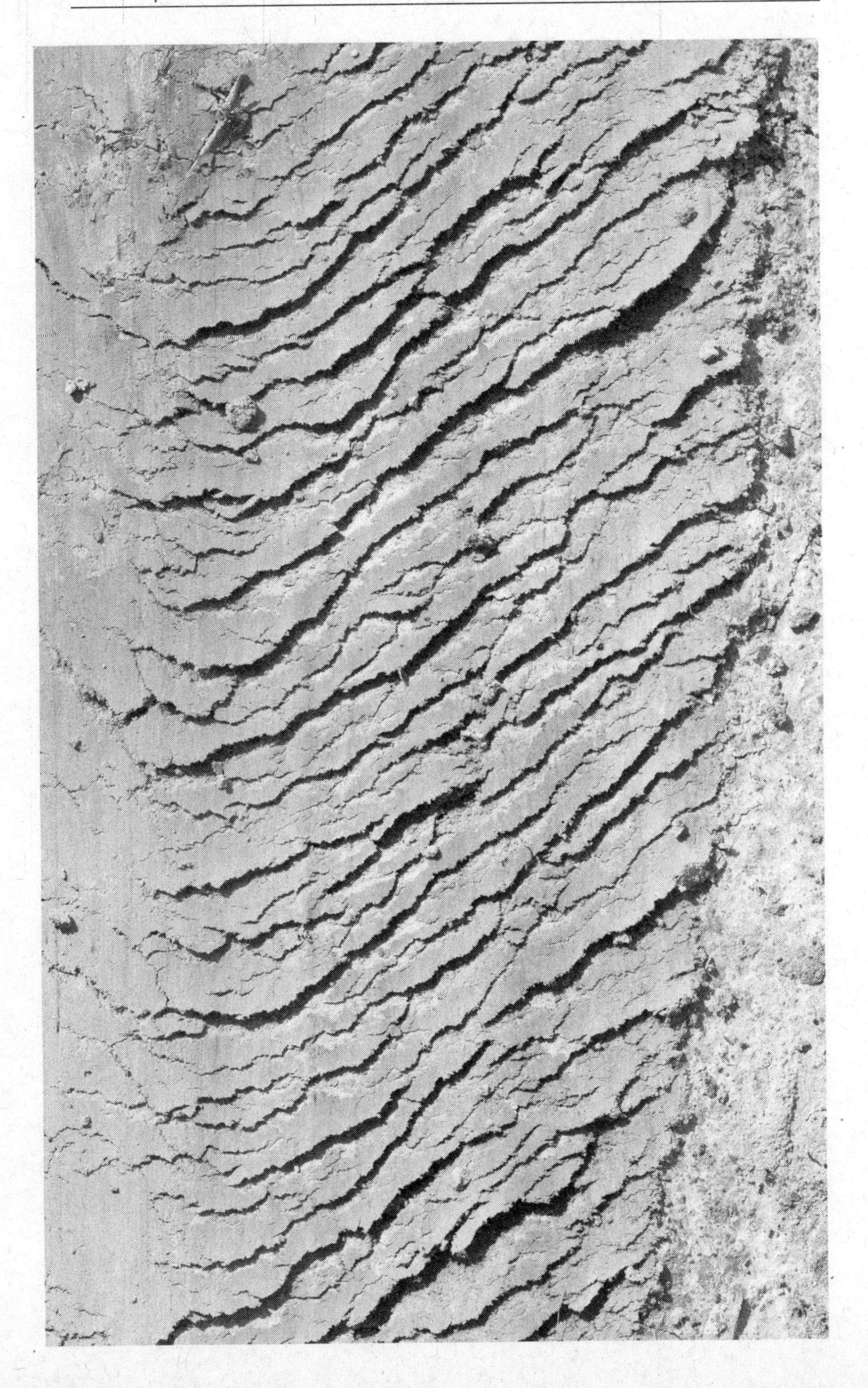

Chapter 2

For All Owner-Built Homes

One aspect that many do not realize before they begin constructing an alternative, owner-built home is that there are plenty of areas where your alternative material of choice will not be used. The difference between building a sod house and building a cordwood house is certainly noticeable from the outside, but there are many systems within these houses that can be strikingly similar.

Far too often, first-time owner-builders forego thoughtful planning for things like plumbing, heating, and electricity; the result of skimping during the planning stages is often increased costs and ramped-up frustration in an already challenging process. Building a home is no small endeavor, but thinking it through beforehand is one of the wisest steps in the process.

Though every home is different, most owner-builders will at some point deal with the following tasks: finding and preparing a site, plumbing, heating and cooling, electrical wiring, roofing, laying a foundation, and installing insulation. While this book will describe the basic techniques, practices, and designs of these tasks simply and succinctly, it is important to remember that the methods described here are not the only possibilities – just the most likely. Volumes of reference materials have been written about each one of these tasks, and no one method is the best for every situation. Do not be surprised if you come across another way to complete these tasks in another resource.

Choosing Your Site

Few people want their home to be situated in a place they find unattractive, but finding a good site goes beyond looks. The composition of the soil itself will play a role in creating a stable foundation; additionally, the existing contours of the land you build on will have a direct effect on the work involved in making the foundation level and on how much effort you will need to put into directing water flow. Where you build should also be close to a good, reliable water source, or you will have a pretty tough time mixing cement, cob, or earth plaster – not to mention showering or staying hydrated. Sun also plays a role in deciding where your home will be built; this is especially true for alternative builders who want to include passive solar heating, a solar water heater, or solar panels for electricity. Wind is a necessity if you plan on using it to supplement your home's power. Even the existing vegetation can play a role on how your house is constructed.

Because there are so many factors to consider when it comes to finding a site, it is certainly advantageous for owner-builders to have several plots of land to choose from. If you begin thinking of the traits you want your land to have before you purchase it, there is a good chance you will be able to find everything you want on one plot – as long as you take your time and shop around. However, many owner-builders are not in a situation in which they can shop around at all; if the land you plan to build on is already in your possession, you may not be able to find one spot where everything you need will be easy to access. This is fine, too. It might take more work to get land in the condition you want it than it does to purchase land that already has everything you need, but perseverance and flexibility are what alternative building is all about.

There are a few basic factors you should look for in a plot you plan to build on. For many, the most important ones have to do with water. You want your home to be on or near an area with good drainage, and in many cases, the site you are building on will be required by code to "perk" or drain at a specified rate. This might not be an issue if you will not be installing a septic system; if you are building in a remote location; or if the structure you are constructing will be more of a getaway cabin rather than a permanent dwelling. Of course, you will still want clean water for cooking, cleaning, bathing, and drinking, so your site should also have easy access to an underground spring, some sort of fresh water reservoir, or municipal water utility. Many alternative building materials are more susceptible to water damage, so constructing your home in a flood plain – even if flooding is rare – is not the wisest idea.

Owner-builders who want to make use of solar energy or passive solar heating will need a site where sunlight is able to broadly hit the house. In the Northern Hemisphere, homes designed to harvest sunlight have their long, window-laden sides facing south. While the northern side of your house can be backed into a hill or blocked by trees, you will want no such obstructions on your home's south side. Additionally, take a close look at vegetation on your site. Certain kinds of flora, particularly climbing vines, can increase the amount of moisture the exterior of your house is exposed to. Be aware that the kind of vegetation that exists naturally on the site will likely continue to exist after your house is built, even if you do your best to eliminate or control it.

Also be aware of your site's soil composition. Topsoil is useful in some cases, but be especially concerned with the composition of the subsoil. The subsoil can change from one spot on a property to the next – and inadequate subsoil can often still be used if properly bolstered – but if an inspector can find good, stable subsoil on your property, it can generally save you any hassle. Good subsoil is stiff, strong, and untouched. High levels of sand and clay are also acceptable, so long as it is tough and difficult (though not impossible) to dig into; remember, you want your base to be solid, but also not impenetrable.

Planning Your Plumbing

Obviously, there is a certain order to consider when planning the components and systems of your home; believe it or not, plumbing is often the best thing to consider up-front. You can even begin planning your plumbing at the same time or before planning your foundation because, in many cases, some major plumbing components will be installed before the foundation is even formed.

When it comes to plumbing in alternative construction, there are two main options. The first is to fit your home with the same type of plumbing you would find in a typical, conventionally constructed house; the second is to reinvent conventional plumbing. The familiarity of traditional plumbing might seem appealing, but when you couple the flaws in these kinds of systems with the accessibility and convenience of alternative options, it is easy to see that septic tanks and sewers are not always the best choice, particularly for owner-builders.

CASE STUDY: CREATING WATER SYSTEMS THAT WORK

Alan Abrams, Principal Designer and CEO
Abrams Design Build
www.abramsdesignbuild.com
alan@abramsdesignbuild.com
409 Butternut Street NW
Washington, DC 20012

Alan Abrams, designer and alternative builder, knows from experience that yesterday's plumbing model can easily be made more efficient and environmentally friendly with today's technologies and innovations.

"My interest in alternative home building was stirred during the eco-conscious awareness movement of the 1970s," Abrams said.

"I bought a piece of land in New Mexico, right on the bank of the Rio Grande, and I sketched the plan for my first house. I was reading Ken Kern's *The Owner Built HOME,* and that initial sketch, which I did in a tent by the light of a lantern, was done on the flyleaf of the book. It was a tiny, free-form adobe home that was built almost entirely by my own hands."

CASE STUDY: CREATING WATER SYSTEMS THAT WORK

"Today, I specialize in residential remodeling ... I incorporate the best practices for energy efficiency, relying on the guidelines laid out in such manuals as the EEBA (Energy and Environmental Building Association)," he continued. "I also incorporate methods for water management, both within the home — like dual flush toilet and kitchen sink foot controls — as well as outside storm water management, like green roofs, porous pavement, and rain gardens."

By using alternative water sources and eco-friendly water management practices, Abrams is helping to change the way builders see not only bathrooms and kitchens, but whole-house designs, as well. By making sustainability a priority in his remodeling designs, he sets a precedent for those around him.

"I devote at least a quarter of my time to promoting green building and educating the public in this area," Abrams said. "I've worked on numerous committees at the local and national level, promoting green building and working on policy. Five years ago, I founded my company to practice the pursuit of sustainability [and] train my staff in green building practices."

As alternative building techniques become more popular and more companies like Abrams' pop up around the country, sustainable water management will likely become the goal of more homeowners. Hopefully in the future, the laws of supply and demand will make the choice to go green as affordable as it is essential.

Plumbing Basics

It has been estimated that the average household uses from 185 to 290 gallons of water every day. About 20 percent of the water is used for laundry, 60 percent used for the toilet and shower, and the remainder used in the kitchen and other indoor areas. In a conventionally built home connected to a water utility, water is brought into a house through a pipe, and waste water leaves a home via a sewer and is brought to a filtration facility to remove the waste. But because you are building your own house, however, you may have the option of making your property self-sufficient when it comes to water usage.

The first step in establishing your plumbing independence is to find a water source all your own. This can mean drilling a well, tapping a nearby pond or reservoir, or even collecting rainwater. In very dry areas, this part of the process can be difficult, indeed; in areas where lakes, rivers, and streams abound, the task is generally easier. But because most of us simply do not have the tools or experience to know how much water we will need, how often our water source will replenish itself, or where to find water underground, this is one situation where calling in a professional can be a lifesaver. Water is a basic need, and one you do not want your home to be without; in fact, having a viable water source is so important that if one does not currently exist where you plan to build, it is probably a good idea to choose another site.

In the same way that a utility company will pipe water into your home, a self-contained, environmentally friendly plumbing system begins with a pipe or hose that leads from your private water source into your house. If you hire a professional to drill a well or tap a nearby reservoir, it is likely that the cost of running the supply line to your house will be included in the bill.

Once it enters the house, you will need to consider a few different options before you go any further. If your water source is not on a higher elevation than your home, gravity will not provide you with adequate water pressure. Here, your supply line will first need to be run to a pressurized water tank in order for water to be distributed throughout the house. Though some alternative builders are seeking a more rugged, rustic lifestyle, most consider having hot, running water a necessity. Hot water can be created in several ways, which will be described in more detail in the "Water Heating Options" section, but if your household is to have heated water, this is the stage when it happens. From the pressurized water tank, you will run two lines out. The first is your cold line, which can be connected directly to your dishwasher, washing machine, tub, shower, and sink. The second line coming from the pressurized tank will go to a water heater before it arrives at your appliances and fixtures.

In many areas of the country, there is another step between the water that enters the house and the water that comes out of the faucet. Water that contains high levels of mineral deposits is called hard water, and when your home's water comes from a hard water well, it is a good idea to include a water softener in your plumbing system to reduce mineral content. Though it might not seem like a big deal at first, hard water can deposit its excess mineral content on the inside of your pipes; when these deposits are allowed to build up over a period of years, they can cause water pressure problems that can take be costly and frustrating to fix. When hard water is present, both hot and cold water should run through the water softener before the lines are run through the rest of the house.

While metal (specifically copper and cast iron) has traditionally been used as piping, plastic PVC plumbing is becoming a favorite among owner-builders. Though it was once seen as simply a cheap alternative to metal, plastic piping is not only less expensive, but easier to work with. Its flexibility allows first-time builders a little leeway when it comes to fitting piping through walls and tight spaces; just as important is the fact that you do not need special tools to cut or join plastic piping. If you already have experience installing metal piping and feel comfortable with the process, there is certainly nothing wrong with using your skills in an alternative house; however, many have found the convenience and ease of installing plastic piping to be a perfect fit for first time and inexperienced builders, as well as builders on a budget. If you have never even touched piping of any sort before, take a trip to your local hardware store or home improvement center and ask for a demonstration of PVC piping; once you see how simple it is to piece together, you may have a better idea of why so many first time builders swear by it. While there may be some inherent downsides to PVC piping —it is the least durable with the shortest lifetime — the negatives generally do not overshadow its many benefits.

It might seem intimidating if you have never done it before, but installing piping between your water heater or pressurized cold water tank is basically a matter of measuring lengths of pipe, adding joints where straight lines are not feasible, and attaching the pipes to the appropriate fixtures. If you do decide to go with plastic piping, cutting the pipes to the desired length can be done with a handsaw, and joining the pipes together will require no welding. Far from rocket science, many first-time owner-builders quickly become well-versed in this task. Even better, once you have installed your own piping, there is a good chance that you will be able to fix many of the plumbing problems that arise over the lifetime of your home without professional assistance.

In some cases, the size of the piping you use can have a significant effect on the performance of your plumbing. For the pipes that lead from your hot- or cold-water tank to an appliance or fixture, a one-half inch pipe is often adequate and will be less expensive to purchase than the three-fourths inch pipe that is more common in new construction. But in general, upping the diameter of the pipes leading to your fixtures and faucets will reduce problems with water pressure as well as problems associated with mineral deposits in hard water situations.

Once you have directed your water inside the home and attached a system to all of your sinks, tubs, showers, and appliances, it is time to plan how it will leave the house. The pipes used for drains and toilet outflow are much larger than the pipes that supply water to your faucets and appliances because they frequently transport more than just water. The largest pipe – generally about 4 inches in diameter – will, of course, be the one that is responsible for carrying all the waste water and any debris that waste water contains away from the house; it should be set on an incline so gravity can easily flush the pipe free of water and debris. When planning conventional plumbing, this largest pipe will lead to your septic system. Here is where the big differences between conventional plumbing and eco-friendly plumbing really start to branch out!

A Note on Plumbing Set Underneath a Concrete Slab Foundation

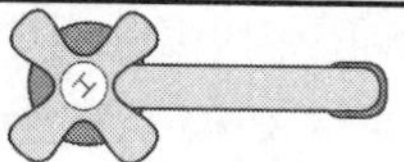

One thing that owner-builders should be aware of is the practice of installing plumbing beneath a slab foundation. In some homes, the large outflow pipe is actually set underneath a poured concrete foundation to reduce the chance of the pipes shifting over time. The outflow pipe leading away from a ground-level toilet may also be set beneath the slab. Some find that under-slab plumbing is a convenient and effective method, but on the downside, if something does go wrong, you will have to get through the slab in order to fix the problem, which can be pretty strenuous and time-consuming. The fact is, however, that installing plumbing components underneath a slab is a relatively common practice among alternative builders, and few have any complaints — even decades down the road.

Installing a Septic Tank

If you choose to hook up to a municipal water works and sewage system, professionals must connect you with the city's water grid. However, most owner-builders who choose to construct an alternatively built home shy away from this option on general principle. For us, the more common choice when it comes to waste water is whether or not to install a conventional septic system.

In conventionally constructed homes that are not hooked up to a municipal sewage line, waste water is almost always handled by a septic tank. A septic tank is basically a large plastic or concrete box buried in the ground that all of the home's waste water is piped into. In a conventional septic system, all the waste water produced in the home – whether from a shower, sink, toilet, or appliance – is fed into the septic tank. The solid waste that enters the tank settles to the bottom where microorganisms eat it; periodically, microorganisms can be added to your septic system to increase the efficiency of this process.

The microorganisms break down your home's waste significantly and are good at their job, but there is a by product of this process: gas. If you examine the plumbing in most conventionally constructed homes, you may notice a curved pipe called a "p-trap" that leads away from sinks, bathtubs, and other places where water drains – the toilet has one, too, but it is harder to spot underneath all that porcelain. This p-trap is handy when something slides down the drain that you wish to retrieve, but another important function served by this handy contraption is providing a constant water barrier that blocks the odored gas produced in your septic system from entering your home. Since the gas has to escape one way or another, the outgoing plumbing in your home should contain an open pipe that extends all the way to your roof, – called a venting pipe – where the gas can be released without excessive odors.

When you use the toilet, run a dishwasher, or take a shower, the water flows down the drain and out into the septic tank; because new water enters your tank many times a day, it stands to reason that old water must somehow exit several times a day as well. The liquid that builds up in the septic tank is allowed to seep out slowly into the ground through a series of perforated pipes that are generally set in trenches filled with gravel. Together, the perforated pipes and trenches are known as a leaching field. The water released into the leaching field is once again fed upon by microorganisms; it is further filtered and purified through the soil and mixed with water from precipitation before it re-joins clean water underground.

A professional often installs the septic tank due to its large size, and this is particularly true of concrete tanks. However, once you know the spot where the tank will go, a handy do-it-yourselfer should have little problem creating his or her own leaching field and connecting it to the tank. Digging the trenches and filling them with gravel is laborious, but it is do-able for any motivated owner-builder. Creating perforated pipe with plastic plumbing can easily be done with PVC piping and a common power drill. The number of perforated pipes and the size of your leaching field will vary depending on your soil content; soil that has a high, dense clay content will absorb water the slowest and will require the largest leaching field.

The standard leaching field will use a series of four trenches that are each about five-feet deep and two-feet wide. The bottom of each trench should be filled with about two feet of gravel, upon which the perforated four-inch-diameter pipe will rest. Once the pipe is in place and connected to the outflow of the septic tank, soil covers both the tank and the pipe. The leaching field should be placed at a lower elevation than the tank itself so that gravity can move the water from the tank to the field without the help of any electrical or other supplementary power.

Water System Options and Alternatives

For many people, the idea of taking perfectly good water, combining it with waste, then sending it off to be separated seems a bit silly when it is a fact that feces and water never have to be joined in the first place. If you plan your plumbing system properly, you can greatly reduce or even eliminate the "black water" that results when feces and water mix and replace it with a mixture of gray water – waste water that is not good to drink, but is not environmentally hazardous – and compost.

Something we often do not consider is that the water that arrives into the toilet is the same that comes through the faucets; it is actually drinkable. Once we mix it with our waste, however, we are polluting it to the point where it must be separated in order to become safe again. Many people feel this is a waste of energy and an unnecessary step. Keeping harmful waste away from potable water is a concept that makes sense to many alternative builders, and this idea is not as difficult to accomplish as you might think.

Soak-Aways

A soak-away functions much like that of a leaching field, minus the septic tank and the perforated pipe. Instead of having all your wastewater lead to a septic tank, you can connect just about everything except the toilet to an outflow pipe that leads to a soak-away. This can greatly reduce the amount of fluid your septic tank will deal with over time, and it will also allow gray water to be released back into the water cycle, skipping its mixing with hazardous human waste. If you plan to install a dishwasher or use bleach or other harsh soaps in your sink, a soak-away can also keep these materials from entering your septic tank and killing the microorganisms inside.

A soak-away sounds simple to build in concept, but it will take some serious elbow grease. It is basically a big hole, often about eight feet in diameter and six-feet deep, which is filled with gravel and covered with a thick layer of straw or hay. Your gray water is piped to the soak-away and is allowed to flow through and be filtered by the hay and gravel before it hits the soil. If you decide to have nothing but self-contained, composting toilets in your home, your house will produce no black water; in this scenario, running all of your outflow pipes to a soak-away is a viable option that will eliminate the need of installing a septic tank.

Composting Toilets

Composting toilets, which compost human waste rather than mix it with clean water, are often considered the best option on the market for the environmentally conscious builder. Not only do these fixtures eliminate black water from your system and the environment at-large, they are easier to install and produce a useful end product rather than a toxic one. When human waste enters a composting toilet, microorganisms break the waste down so that it can be utilized for a garden. While some people are repulsed by the idea of using human waste for a home garden, a good composting system both in the toilet and outside in a composting bin breaks down the waste to a usable, more appealing state. This system stops our waste from entering the water system either through public sanitation or through individual septic systems that leach into groundwater.

Far from the conception that many have of composting toilets being no more than indoor outhouses, modern composting toilets do not require the homeowner to bear any overly noxious scents or even deal with recognizable waste, as these units are designed with homeowner comfort in mind. With the correctly sized unit for your household, the most intense step you will have to do yourself is remove a bin or tray of ready-to-use compost from your home and out to your garden. The size and style of a composting toilet will determine how often this should be done – some larger systems will not require much attention for months at a time. Not only will you be helping the environment and your tomato plants, you will save on water costs when compared to dealing with a utility or installing a septic tank. You may even be able to receive government incentives to help ease the burden of installation costs.

Using Rain Water

Green builders often use rain catchments in addition to accessing another type of water source because storing rainwater that drains down gutters also serves the purpose of collecting water that might otherwise be absorbed back into the ground near the foundation. Should the water be used for drinking, the roof from which it drains should be checked for any chemical that could be absorbed from the roof (or any sealant/paint that covers the roof) and make the water toxic. The water should also be filtered. Dependable filtering systems that extract all organic and inorganic toxins are quite expensive, so such a decision requires careful consideration. However, rain catchment systems are excellent for providing water for gardens and other non-potable needs.

If you are looking to install rain catchments only to reduce the amount of water that hits your home's foundation or supplement your gardening water supply, a few rain barrels should be able to do the trick. When simply positioned under a downspout, nearly any large container can serve this purpose. However, purchasing or constructing a rain barrel that has a spigot on the bottom and a lid that can be tightly sealed and tightly fitted to your downspout is a generally good idea. Not only will these two simple barrel additions make attaching a hose easier, but the tight seal will reduce the likelihood of mosquitoes breeding in your barrels.

Keeping Your Water High Quality

The importance of water quality is yet another integral feature of a sustainable home. We cannot live without good-quality water, which is rapidly becoming scarcer. To help keep the water that leaves the home as clean as possible, resist using cleaning and other products that are toxic; many non-toxic products containing natural cleaning agents, such as orange peel oil, are readily available and are labeled as such. Using biodegradable products reduces the toxic load already coursing underground, which is quite considerable as-is. Be careful about what gets washed down the drain because, while it is a good idea to test the water whatever the source, it also helps to know that what we put into our own bodies ends up in some manner or another back into the water supply. Coffee and medications, along with pollutants in the air and in the ground, seep through into groundwater that, though often filtered through an aquifer, might still carry a variety of toxins that people would not knowingly use for drinking or for showering.

Drilling a Well

A well can be an ideal situation for those looking to create a sustainable, off-the-grid house. Most choose to use an electric pump to obtain water from their well into their home, but those who truly want to use as few resources or save as much money as possible have successfully employed hand pumps and even bicycles. Even if you do use an electric well pump, it is not a bad idea to include a hand pump into the design to ensure you can gain easy access to water during a power outage.

Just to be clear: Do not plan to drill your own well. Unless you are or were employed as a well driller and have access to the tools needed to do the job, you cannot complete this task on your own. Professional well drilling can be expensive, but there is simply no other way for you to get this job done right. Costs can vary from one situation to the next, as many factors such as water depth, soil composition, and local cost of living can all influence price, but in most cases, do not be surprised if estimates to have a professional drill a well and run a supply line to your house are in the $5,000 to $10,000 range. In some situations, the costs may be even higher.

Connecting to a Utility

If you live in a city and can successfully obtain all permits needed to build your own home, consider connecting to a municipal water company. Because many cities are short on undeveloped land to begin with, and because building codes in cities are far stricter than they are in most rural areas, connecting to a utility company often makes better sense than installing a septic system and drilling a well in an urban or suburban area. Remember, creating a house you can live in means complying with all the laws of the land, and in a city, this often means paying for licensed professionals instead of doing the job yourself. In many cases, your options in an urban area will be more limited than they would be in the country, and for many owner-builders, the lower cost and higher number of construction options are the most compelling reasons to build on rural land.

Water Heating Options

Hot water may not be an absolute necessity, but it may be hard to live without. Each water-heating method has varied pros and cons. Solar water heaters harness the power of the sun to provide your home with hot water. If you have built in an area where direct sunlight is plentiful, this option is not only the greenest, but also likely the cheapest in the long run. When mounted in or on your roof, water that is heated by the sun and stored in a tank will be at your disposal and pressurized by gravity. Some solar water heaters that are designed for colder climates use the sun to heat anti-freeze instead of directly heating the water itself. The anti-freeze is then passed through a heat exchanger to provide heat for water in your home. This option is cost-effective, but in low sun exposure and very cold areas, it may need to be used in combination with conventional water-heating techniques.

Conventional hot water tanks are another viable option, though most green builders will tell you that they are far from optimal. Because hot water tanks only hold a finite amount of water, you must ensure you use a tank that is large enough — the largest residential ones can hold more than 100 gallons — to provide hot water for all the tasks your household will require in an average day. Many have found that the size of their hot water tank is inadequate and have paid for this miscalculation in cold showers. Another problem with traditional hot water tanks lies in their energy consumption. The water in the tank must be kept hot enough to use at a moment's notice, which means that your electricity or fuel will be used throughout the day, even when you have no need of hot water. This in turn leads to wasted energy and higher fuel consumption. Hot water tanks heated by electricity tend to be less efficient than those heated by natural gas or propane, and an electric unit will typically require more storage capacity than a gas unit for the same size household, as well.

Instantaneous or tank-less water heaters warm up water only when it is necessary. When you turn on your hot faucet or set your washing machine to a warm setting, the heater kicks in; when you are not using water or are only using cold water, the instantaneous water heater does nothing. The result: Hot water when you want it, for as long as you need it, without any wasted fuel or energy. You can purchase both large, whole-house tank-less units, or you can fit small units designed to fulfill the needs of a single room. Depending on your level of water consumption and the size of your house, one option may be less expensive or more effective than the other, so shop around thoroughly. Tank-less water heaters may indeed be a better, more efficient fit for your home, but the initial cost of purchasing one of these units is also typically higher than a conventional water heater.

Laying Your Foundation

No house can expect a long life if it is not built upon a stable foundation. Creating a solid foundation on which a new house will stand could be considered the most important step in the building process, as a poor foundation can literally bring the house down. Consequently, some owner-builders choose to hire a professional for this job.

If you want to build your own foundation, be aware that the larg-er the house is to be, the more meticulous and exact you must be when planning it. Those building a small structure or guest house can worry less about getting every measurement perfect, but those planning on constructing a permanent dwelling suitable for housing several family members need to be willing to take the necessary time and put in the required effort to do this job precisely. While taking shortcuts might be passable some facets of home building, an insufficient or hurried foundation can have dire consequences in the long run. And if foundation problems are not spotted until the house is finished, they can be difficult and expensive to repair.

Making a Big Rectangle

Your middle-school geometry class is about to pay off. In most cases, creating a space to build will involve drawing a big rectangle on the ground to know where your foundation — whatever form it may take — will be formed. To do this, you will need a few simple tools: a line level — a small level designed to attach to a string — three metal or wooden stakes, a lengthy tape measure, and a big ball of string.

Unless you are building a round house, your exterior walls will need to meet at right angles to provide stability. Creating a big rectangle is not hard, but it can be tricky if you do not know what to do. Here is a rather simple way to do it:

Once you have got the actual dimensions of your house drawn, you must transfer those from paper to the ground. To do this, we use the Pythagorean theorem, altered here for our purposes: width x width + length x length = diagonal x diagonal.

- Say the width of your house-to-be is 30 feet and the length is 40 feet. In this example, width x width (30 x 30 = 900) and length x length (40 x 40 = 1,600), 900 + 1600 = 2500, which also equals diagonal x diagonal.

- Take the square root of 2,500 and you get 50, which is the length of the diagonal.

- Use the tape measure to cut yourself three lengths of string (30 feet, 40 feet, and 50 feet) and stretch them taut to form a perfect right triangle. Mark each corner.

- Keeping the diagonal string and the points where the other two strings meet the diagonal in the same position, you can find the other corner of your perfect rectangle.

On a flat surface, this job is now complete, but because it is often necessary to form your big rectangle before the ground is made level, this is where the stakes and line levels come in.

- Attach the strings to the stakes and the levels to the strings. Pull the strings taut just as before, but move the strings up and down on the stakes until the bubble says the strings are level.

- Once all strings are taut and all bubbles indicate that they are level, you can now mark the corners with your stakes. Repeat as before to find your fourth rectangle corner.

If you are planning a round house, it is very easy once the ground is level, but rather difficult when it is not.

- To make a perfect circle on level ground, take a string that is half the diameter of your intended floor space and attach one end to a stick.

- Drive the stick into the intended center of your house and stretch the string taut. As long as the string stays taut and level, you can only move along the perimeter of your house's floor plan.

Know Your Subsoil

Subsoil is the dirt that lies underneath the topsoil. Topsoil is simply not strong or reliable enough to build upon, so you must get underneath it to find ground on which to build. Most topsoil only permeates a foot or less under the surface, but it is important that the subsoil you uncover is completely undisturbed. In the United States, where so much land has been altered, you may have to dig several feet into the earth to find undisturbed ground. While a foundation is the structure that your house is built on, the subsoil is the base from which your foundation will draw its stability. Thus, it is in your best interests to find a site where the subsoil is both suitable and relatively easy to access.

Soil that is suitable to build on will have a few identifying characteristics: It will be solid and difficult to compact; it will be continuous over the entire area of the intended structure; and in colder areas, it will be below the frost line. If you live in an area where freezing temperatures are inevitable, this last characteristic is especially important. Your local government should have information on how deep you must dig to get below the frost line, and because frost heave can crack even the best-laid foundations, it is recommended that you procure this information.

When you are digging for suitable ground, there are many things that you can uncover. Finding solid rock or firm clay beneath the soil will let you know that you have found a place where the weight of your house-to-be will be easily borne. Uncovering weak, easily disturbed subsoil means that you have to either dig deeper, design with weak soil in mind, or find another spot. If you find extremely wet, muddy subsoil, seriously consider picking another site for your structure. Ideally, once you hit undisturbed ground, you will find dry, hard earth that is difficult to pierce more than a one-half inch or so with anything relatively blunt, like a piece of rebar. However, if your subsoil is on the soft side, you can increase the stability of the planned structure by creating a thicker, wider, and stronger foundation.

Foundation Footings

In most cases, a "footing" is a hole dug in undisturbed ground that is filled with concrete. Footings can be poured individually to provide stability for piers or posts, or they can make a continuous concrete base that traces the perimeter of the structure that is to be built upon it. Whether you have individual footings to support posts and piers, or a continuous – "monolithic" – footing to support the entire structure's perimeter, the main purpose is the same: to provide an anchored, stable spot on which to build.

In a rectangular structure, pier footings are generally placed underneath where the corners of the structure will be; the corners are an excellent place for the piers to bear the weight of the house that will be built on top of them. Footings will also typically be placed underneath any location where a load-bearing post in a post-and-beam framework is desired. Constructing a continuous footing that traces the entire perimeter of the house provides extra strength and stability at every point where an exterior wall will be erected, instead of just at the corners and under posts. In many modern constructions, a monolithic footer is the base.

Unfortunately, laying a continuous footing that traces the perimeter of the house will use a lot more concrete — and cost more money — than setting footings at the corners of a structure, but the strength and stability of a monolithic footing design is unmatched. Some alternative builders prefer to construct pier-type footings because concrete requires a large amount of energy to produce, mix, and transport. In buildings where walls are particularly heavy, like cordwood, a monolithic footing might be worth the additional cost and energy expenditure. Continuous footings around the perimeter of a structure can also be beneficial when building round or curved houses where there are no discernible corners.

What kind of dimensions should you have for your footings? This is a difficult question to answer, as the minimum dimensions for a stable house can vary dramatically from one structure to the next, depending on the height of the building, the width and weight of its exterior walls, the weight of the roof, and the soil composition in the area. Some small structures have lasted decades with footings only a few inches deep, formed from bricks or stones, but a two-story cordwood house with a living roof would hardly be called "stable" if built on such footings. There is not one answer when it comes to minimum dimensions for your footings. However, while no one can tell you exactly what kind of footings you should choose and what their dimensions should be, it is possible to gain an idea of what will work in your situation by taking a look at projects that have been successful in the past.

In his book *Complete Book of Cordwood Masonry House Building*, Rob Roy advocates continuous footings between 8- and 9-inches deep and between 16- and 32-inches wide; the larger width dimensions are appropriate when the structure has two stories, a heavy earth roof, or in areas where subsoil is less stable. This kind of foundation would be large, but because Roy is building structures with wide walls that are extremely heavy, the type of strength needed to ensure a stable base cannot be achieved with a lesser footing. In Clarke Snell and Tim Callahan's *Building Green: A Complete Guide to How-To Alternative Building Methods*, their "little building" — a hybrid of cob, cordwood, and straw bale construction — is set on four footings that are each 24-inches wide, 24-inches long, and 12-inches high and are positioned at the corners of the structure; these are then connected to each other by four gravel trenches that line the perimeter of the structure.

Obviously, the latter scenario uses far less concrete than the former by replacing much of the concrete used on the perimeter by gravel, but the larger, continuous concrete footing suggested by Roy is suitable for bearing a far greater load. Snell and Callahan's concrete footings and gravel trenches are excellent choices for their situation, but the dimensions of the cottage they are constructing are smaller than those in an average house. When a larger, heavier structure is the ultimate goal, there is no harm in erring on the side of caution.

Gravel Trenches

In Snell and Callahan's little building, corner footings and gravel trenches that trace the perimeter of the structure provide the foundation. This method is popular in the owner-builder world for several reasons: It is affordable, is energy-efficient because it minimizes the use of concrete, and is eco-friendly because it does not disturb the land as much as some other options do. Its drawback: Gravel is not as structurally sound as concrete. When building very heavy walls or constructing multiple stories, gravel trench foundations may not be the best choice. If you are constructing a fairly small, one-story dwelling, however, gravel trenches can create an effective base.

A gravel trench foundation begins with a trench. This can be done by hand with shovels if you have the time and energy, but most will find that the use of a backhoe makes the job go by more quickly. Your trench should be a minimum of 16-inches wide and should be dug at least 4 inches below the frost line in your area; wider trenches are needed for heavier buildings and for less stable subsoil. The trench should follow the intended perimeter of the house, and the center of the trench and the center of your intended walls should align. The trench should also slope slightly downward and away from the house. If there is a natural slope to your property, the trench bottom should slope in the same direction; if there is no natural slope to your property, the trench bottom should be designed for water to flow toward the area where your septic tank and/or soak away will be placed.

After your trench is at the intended width and depth, and your subsoil is tamped down and stable, install a drainpipe. This is simply a continuous piece of heavy-duty, perforated, 4-inch flexible plastic tubing that will line the bottom of your entire trench. If your soil is heavy on sand and silt, this tube can be wrapped in a fabric sleeve to keep dirt from clogging up the perforations. This is probably a good precaution for any soil, as it will not take long to do or cost much. Fill the bottom of the trench with about 4 inches of gravel, then lay in your drainage tube; since the trench bottom is already slightly sloped with the natural grade of the property or toward the septic tank or soak-away area, gravity will carry any water that enters the tube away from your property.

After your drainage tube is in place, you can construct your footings underneath where posts will be placed. When your footings are set and your drain is set, fill in the rest of the trench with gravel in 1.5-inch increments, making sure to stop and tamp the gravel down after each level is added until your gravel and the surrounding subsoil are on the same level.

Once this process is complete, you can build stem walls on top of your gravel trenches, or build forms that line the entire outside perimeter of your foundation and pour a continuous concrete slab that covers the trenches as well as the entire floor space of your house to be.

Concrete Slab Foundations

A 4-inch-thick concrete slab foundation can create a solid, stable base for just about any structure you could ever think to build yourself. Not only does a slab foundation provide structural stability for the shell of your house, it also provides a good base to anchor any interior framing you may choose to include. Pouring a slab is generally done in either one or two steps. For smaller structures, you can pour the footings and the slab at the same time; with larger buildings, you should pour the footings one day and the slab on another. Rob Roy suggests that this method be used on houses with dimensions longer than 40 feet in any one direction.

For the owner-builder who plans on constructing every single aspect of his or her home without professional assistance, pouring a slab may not be feasible. Unless the slab in question is for a shed, very small cottage, or other modestly sized building, you will likely need to call in a professional concrete service due to the sheer volume of concrete that will need to be poured. Just because you have to call in a big truck to pour the concrete does not mean you will not have any work to do yourself, however; it just means that you can focus on evening out the surface of the concrete as it is poured rather than expending your energy trying to mix enough small batches of concrete to create the slab.

While concrete is a durable and tough material, to make a slab even more stable, it is common practice to reinforce the concrete's already significant strength by binding it with wire mesh. By placing wire mesh inside your slab, you are supplementing the bond concrete creates to itself with the strength of metal. This supplement will help keep your concrete from cracking and can even hold your foundation together if cracking does eventually occur. Look for a 10-gauge wire mesh that runs in a 6- by 6-inch grid.

Another easy and effective method of reinforcing concrete is mixing fiberglass into the concrete. These tough fibers become distributed throughout the concrete and resist any contraction. If you order concrete delivered from a ready-mix company, the company can add this before they leave the plant, and it will be ready to pour at arrival. It does leave little bits of fiberglass poking up in places, so this is not an ideal option if the pad is intended to be the final floor.

One thing to remember about concrete as a material: It does not drain well. Because of this, anyone planning on pouring a concrete slab foundation should be concerned about drainage underneath the slab. To combat the effects of heavy precipitation and moisture, concrete slabs are often poured not directly on the subsoil, but on a pad of coarse sand or gravel. This pad – generally between 15- and 18- inches deep and extending 3 feet past the intended perimeter of the structure – will ensure a uniform, equal settling of the slab and provide a base that water can easily pass through, keeping moisture away from the slab. Most experts also recommend placing a waterproof membrane (a big sheet of plastic) on top of the pad before the slab is poured. This membrane will provide additional moisture protection for the slab on top of it.

Under-floor drains set in the sand or gravel pad can also be beneficial in reducing problems with moisture beneath the slab. Because they are inexpensive and easy to install, it is recommended that those pouring a slab foundation include under-floor drains in their designs. All that is needed to create an under-floor drain is flexible, perforated plastic tubing – 4 inches in diameter should be sufficient – and some nylon or fiberglass material to be wrapped around it. The nylon or fiberglass sock wrapped around your tubing will prevent silt from plugging up the holes in your tubing. The process entails burying the fabric-wrapped tubing in the sand or gravel pad until only the very top of it is visible. By angling this tubing away from your house and directing it toward a soak-away, you can enhance your foundation's drainage capability.

Homeowners in cold areas should also consider supplementing their slab foundations or footings with Styrofoam™ insulation. Concrete transfers heat rather readily, and one-inch thick Styrofoam boards laid underneath the poured concrete can significantly reduce heat loss during a cold winter. In areas like New England or North Dakota, where winters are particularly harsh, consider placing Styrofoam on the outer edge of the footings – like an insulated baseboard that goes all around the foundation – as well as laying a continuous layer of Styrofoam that lines the entire underside of the slab and footings. In the south and southwest, you might skip the insulation altogether. It is important to note that not every Styrofoam board is tough enough to hold up under tons of concrete. Rob Roy recommends only one: Dow Blue Styrofoam.

Basements

Unless there is a specific reason for needing a basement, many experienced alternative builders advise against the construction of one. Not only are basements often too difficult for many first-time builders to construct without time-consuming and costly mistakes, but the components of a home that are often housed in basements like water heaters, furnaces, and plumbing lines may just as easily be kept on your first floor.

Basements and other earth-sheltered spaces do have certain advantages when it comes to maintaining constant temperatures, and materials like insulated concrete forms can make their construction easier, but those owner-builders who absolutely must have a basement should enlist the help of a professional to plan and help build the space. Constructing a basement is a complicated process, and one that should not be taken lightly; even a minor error can cost thousands to fix. Having a professional do the job is one way to ensure successful results, but because the cost of most professional basement constructions will make up a large fraction of the total cost of many alternative homes — about $20,000 or $30,000 for low-end basement constructions — this is one operation that is often more trouble and money than it is worth.

Crawlspaces

In conventional construction, a crawlspace is an area underneath the ground floor of a house that is meant to provide easy access to pipes and other elements of the home. Often built mostly or entirely above grade, a crawlspace is like a miniature basement that is too short to walk upright in. Extending the piers or the stem walls up from the footings in a foundation will create a suitable crawlspace.

In alternative construction, crawlspaces are not always a good option — at least, not in the conventional sense. In many alternative homes, the exterior walls are simply too heavy and thick to feasibly be supported by extended piers or stem walls. However, it is possible to create a pseudo-crawlspace in several areas of the house that will allow easy access to pipes or wiring by raising the floor in those areas. It will take extra work and some forethought in design, but creating a pseudo-crawlspace to house some components of your home systems could save you trouble down the road if access to those systems becomes necessary.

Creating a pseudo crawlspace can be a part of installing your permanent flooring. Instead of installing your floor directly on top of the foundation, you can raise your floor just high enough to accommodate plumbing and wiring. First, you will need to install some joists to support your floor; this is accomplished by installing a series of 2x8s, which are attached both to the walls of the room and anchored to the foundation. This will allow you to install a plywood base to the 2x8s that will give you a subfloor suitable to install carpet or another flooring material on top. The components underneath will then be easier to access by pulling up the surface flooring material and the plywood. If you plan on taking this route, it is important to remember that this will raise your floor by eight inches. It is necessary to plan this step early so your ceilings will be high enough to allow easy walking; it is more difficult and costly to raise your ceilings later than it is to build them higher in the first place.

Pouring Concrete

Effectively and efficiently pouring concrete is not a difficult job if you know the proper steps. Concrete must be poured into a form so that it can solidify and cure to become the shape you desire. For some jobs, like certain types of foundation footings, the form you pour your concrete into is the trench or hole you dig in the ground; in most cases, however, the form is something you must construct. This is often done with simple lumber that is nailed or screwed together and anchored in position with wooden spikes or metal bars hammered into the ground. When pouring piers, you may be able to find the cylindrical, cardboard forms that are used in commercial construction. These cardboard forms are convenient and can simply be cut away with a utility knife after the concrete has cured, leaving you with a sturdy, no-fuss cylindrical concrete pier.

If choosing to embed mesh in the middle of the concrete to strengthen it, you will want to lift this mesh up off the ground so it sets within, rather than on the bottom of, the concrete. Metal poles called rebar, rather than wire mesh, generally strengthen piers and footers. Rebar can generally be fixed in the desired position before the pour, so do not worry about having someone move it into place while the pour is happening.

Once the liquid concrete is successfully poured and in its form, it must be made level. The first step in leveling the concrete is called screeding. For do-it-yourselfers, this process is done by pulling a long, straight, sometimes weighted board over the freshly poured concrete to flatten it out. The screed board sweeps away bumps but does not do anything for pits and depressions, so it is good to have someone with a shovel on standby to fill these in. The more continuous the surface is left after screeding, the easier it will be later, so take your time and get the surface as flat and even as possible here. If you have a large enough crew – you should have at least five or six people on site when pouring large expanses of concrete, and hopefully one or two will have done this before – you can screed one section as another is being poured.

Next, the wet concrete must be floated; this process is designed to pull water from inside the concrete to the slab surface and is done with a bull float. The bull float is a big metal trowel that is attached to a long enough pole that the operator can reach every section of concrete without having to stand in it. The large metal surface of the float is laid flat on the concrete and then agitated enough to bring water to the surface, but not enough to create deep grooves or cuts in the concrete surface.

The last step in concrete leveling is troweling. For small areas, you can get away with using a hand trowel to finish the surface and make it completely flat. On large expanses, using a power trowel is more efficient. A power trowel is a specialized piece of machinery that requires experience to use safely and properly; if you have not used one before, now is not the time to learn.

The process of liquid concrete becoming solid concrete is called curing. In most scenarios, it is important that concrete is not exposed to extra moisture while it is curing. In dry climates and on very hot days, you should add water to the curing concrete's surface to keep it from solidifying too quickly; concrete that cures rapidly can often develop cracks. Once the sun goes down, the concrete will be able to cure at a more moderate pace, so it is necessary to cover the concrete with a plastic sheet or tarp to keep it protected. The plastic should be anchored with weights to keep from blowing away, but make sure the weights are not on the concrete itself.

Electrical Wiring

The information put forth in this section is meant to describe the basic principles behind electrical wiring and to help those who have never done the job before understand those principles. If you plan on doing your own wiring from beginning to end, however, it is essential that you at least consult with a professional. This chapter is not meant to be a comprehensive manual on every wiring option available or a detailed explanation of the wiring codes in your area, but it will equip you with the basics so that when you do ask a professional, you will already know the fundamentals.

Understanding Basic Wiring

The principles behind electrical wiring can be explained rather easily. In a conventional home, electrical power enters the house through a single large cable that is installed by the power company. This cable is then attached to a breaker box where it is split. From the breaker box run many smaller cables, which are placed behind the walls of your home; these smaller cables carry different varying loads of electricity and provide power to different parts of the house. In each room, these cables are then attached to light fixtures, electrical outlets, and switches of their appropriate electrical capacity.

Wiring in an alternatively constructed home is usually similar in concept to that of a conventional home, and is sometimes exactly the same. However, because so many alternative owner-builders are interested in reducing not just the environmental impact of the building process but the impact of their energy consumption over the lifetime of the house, there are certain differences that often appear in such homes. It is up to you or the electrician you hire to plan just how similar to a conventional home your house's wiring will be. This planning must be done early, as it is difficult and expensive to change your wiring design once the shell of the house is constructed.

First, let us tackle the basic wiring techniques that are found in most conventional and many alternative houses. If you find yourself asking questions in your head while you are installing your wiring, you must consult a professional before moving forward. Wiring is the last place where you can afford to make a mistake; if you find yourself unsure at any time in the process, bite the bullet and call a pro, as the consequences of installing wiring improperly can be fatal.

About Voltage

Describing voltage can be confusing. When the power company runs a line into your house, the total voltage of the line is, in most U.S. residences, 240 volts. These 240 volts come in on two different wires of 120 volts each. The outlets and switches placed in various areas of the house will be stamped with a label that says 120 volts, and the devices you plug into your outlets will generally say the same thing. It sounds simple, but problems arise, however, for owner-builders using more than one resource to do their wiring. While one resource might refer to a 120-volt outlet or device, another might refer to a 115-, 110-, or even 125-volt outlet or device and mean the same thing.

The discrepancy here comes from trying to be too precise: By the time electricity runs through all the necessary cable and wiring and reaches an outlet, the overall voltage can decrease due to resistance. The longer the wire path, the more likely it is for resistance to occur. On a substantial path, 120 volts can become 115 or 110 volts, and the experts will often refer to this voltage loss as they discuss wiring. Voltage loss due to resistance is beyond the scope of this book, but knowing that another resource might refer to an outlet as 110, 115, or 125 and still mean 120 may clear up some confusion down the line.

From Pole to Panel

The attaching of the main cable that comes to your home from a utility pole to your home's electrical panel is the first step in wiring your home. In old houses, the electrical panel is a box with many fuses in it; in new construction, these fuses have been replaced with more convenient, safer circuit breakers. The cable coming from the utility pole will consist of three main wires: two hot wires that will actually carry electricity into your home, and a neutral wire that is responsible for outgoing electrical flow. In U.S. residences, each hot wire carries 120 volts to form a 240-volt system. Each of these three wires will have a designated attachment point in your breaker box.

A fourth wire, called the ground wire, will also be connected to your breaker box. This wire is quite important, as it is responsible for carrying electricity away from your system and safely to the ground in the event of surges and spikes. The ground wire can either be connected to metal rods that are driven into the earth or set in your home's foundation, or can, in some cases, be attached to metal plumbing in the ground. In most places in the U.S., building code requires that the neutral wire is attached to the ground at the breaker box.

The wires coming in from the utility pole should not be flowing with electricity when the breaker box is set up, and the electric company can shut off the current so that it is safe to work on them. Grounding is an important safety feature that should be overseen by someone that knows what he or she is doing. In fact, there are many benefits to paying a professional to connect your incoming power line to your breaker box, determine its placement, and set up your ground wire; at the very least, consider consulting a professional directly to have him or her explain exactly how to do this job. Yes, this will be just another cost to add to the pile, but it is one that will ease your mind and save you trouble. Once the box is in place and your system is properly grounded, you will be able to turn off electrical current to individual circuits and safely connect switches, fixtures, and outlets in one section of the home at a time.

Thinking Inside the Box

A wire and cable have distinctions from each other. A wire is basically a string made of metal that is often encased in a colored, plastic, insulating sheath; in most home wiring, that metal is copper. A cable consists of several wires that are bundled in a plastic sheath. When you cut open a cable, there are wires in it; when you cut open a wire, there is nothing but metal. Attaching the appropriate wires will make the connections to individual outlets and switches; getting these wires from the breaker box to the outlets and switches is done with cables.

When your breaker box is set up, look inside it. You will see a series of switches – these are the breakers – that are each labeled with a number. This number tells you the maximum number of amps the cable attached to a particular breaker can deliver. In a simple set-up, these numbers will consist of a couple 40s and 30s, and several 15s and 20s. These numbers describe the amount of amps the switch is meant to handle. Normal light switches will generally be attached to the 15s and normal outlets to the 20s.

You will also notice that some of the breakers are single switches that move individually while others are double switches that move in pairs. These double switches will usually be labeled with a number 30 or higher and are meant to provide power to devices like electric stoves, dryers, and water heaters that need 240, rather than 120 volts, to operate. In these cases, each 30 or higher amp switch will carry the normal 120 volts, but they will combine to send 240 volts to a single appliance, which is why they operate together. The wall switches you will be connecting to the cables leading from a 15 will say "15 amps" on them; the outlets that will be connected to cables leading from a 20 will say "20 amps."

The more amps a cable will be responsible for, the larger in diameter the wires inside that cable must be. This is why the cable that runs from a 15-amp breaker to a 15-amp light switch is smaller than the cable that runs from a 30-amp breaker to the 30-amp outlet your dryer will get plugged into. A cable's diameter is known as its gauge. Note that the smaller the gauge number is, the thicker the cable will be. An 8-gauge cable will be much thicker than a 14-gauge cable and will therefore be more appropriate for a larger electrical load.

Choosing the right size cable is important; in basic wiring, 14-gauge cable is the bare minimum for connecting 15-amp switches — though some homeowners and building codes do away with 14-gauge cables altogether, making 12 the smallest diameter cable you will use — 12-gauge cables are appropriate for connecting 20-amp outlets; 10-gauge cables are for 30-amp outlets; and 8-gauge cables are used for 40-amp connections. These gauge sizes are for copper cable; you should not be running aluminum cable to switches, outlets, or appliances. Aluminum cable does not transfer electricity as efficiently as copper cable, and though used by professionals in certain applications, all wiring done between your breaker box and your outlets and switches should be done with copper.

From the Box to Your Living Room

Bulk cable that you purchase at the hardware store or online will be labeled first with the gauge and then with the number of smaller wires the cable contains (generally either 2 plus ground or 3 plus ground). The ground wire is for safety and directs current, logically, to the ground via the larger ground wire connected to the breaker box. The hot wire, again, is the one that provides the actual electricity to the device. Because electricity flows in a loop, called a circuit, the flow of electricity does not end at the device it is entering; this is where the neutral wire comes in. Much like the outflow drains in your plumbing, the neutral wire carries electricity away from the device; the flow of electricity goes through the neutral wire and back to the breaker, completing the circuit.

For most of your home's wiring, you will be dealing with either 14- or 12-gauge cable that is labeled "two plus ground," which will include one ground wire, one hot wire, and one neutral wire. In some cases, you will deal with a larger gauge cable or cable labeled "three plus ground;" this generally happens when you are hooking up a special-use outlet or a three- or four-way switch. Ranges, dryers, water heaters, and other large electrical devices sometimes require a connection that uses two hot wires instead of one. In such connections, the device will be running on 240 rather than 120 volts; the second hot wire is responsible for carrying this extra voltage. Depending on the number of amps the device will run on, the special use outlet it will plug into may be connected to your breaker box with an 8- or 10-gauge cable.

If you are looking at more than one resource to help you do your wiring, you are bound to notice a problem with the descriptions of how the three basic wires inside a cable are colored. Countless TV shows and movies have made many of us very wary of wire color — a person's fate hangs on the answer to a single question:

"Which one do I cut, the black or the green? — In many cases, the wire coloring goes like this: Black means hot, white or white stripe means neutral, and bare means ground.

There are variations, however, which can be particularly confusing when dealing with an existing system. You may run into ground wires that are encased in green plastic, or hot wires that are red when dealing with 240-volt connections. You, however, are in luck because you are putting in this system from beginning to end. If you purchase all your wire at once or always buy the same kind of wire, you will always know what color of wire means what. If you stay within building code, all your black and red are hot wires; all white is neutral; and all green or bare is ground. Deviating from that could spell trouble for anyone who comes along in the future to work on your system.

Making Connections

It easy to say, "run this kind of cable between here and here," but when it comes time to actually do it, many find they are intimidated. Again, if you are unsure of your ability to do this kind of work, talking to someone experienced can be a lifesaver. If you do not personally know anyone who does wiring for a living, it might even be worth the time to contact a few pros you do not know and discuss the kind of house you are building. It might sound silly, but people — particularly those in the construction industry — find it intriguing to see someone build a house out of straw, tires, or firewood. In many cases, you can end up with a cheap or free tutorial on how to complete some parts of the project simply by telling a curious professional what you are up to.

Whether you have a professional on-site or you are flying solo, the first rule of working with electricity is to ensure the circuit you are working on is turned off. This rule cannot be emphasized enough; the failure to follow this rule can have disastrous results. Shut the current to the circuit you are planning to work on off at the breaker box every time, and always proceed as if the wires you are working with are hot, even if you know they are not. Safety is a vital issue here; do not skimp on it.

Once the current is shut off for the circuit you plan on working on, it is time to connect your switches and outlets with the appropriate gauge wire. You will need a couple of tools for this job: a utility knife, needle nose pliers, a wire stripper, a drill with a Phillips head bit, and a Phillips head screwdriver. You will likely have these tools already, except for the wire stripper. It is worth the expense to purchase this device, though some tutorials tell you that you can get by without it. But if you are wiring an entire house, this is a small device that can be a significant time saver.

Let us start with wiring an outlet. You will need an outlet box, which come in both metal and plastic, but plastic is cheaper and can be used unless code specifically prohibits it. You will also need a length of appropriate cable (12 gauge, 2 plus ground) that is long enough to run along the wall(s) between the breaker box to the outlet; you should add about a foot to this measured length to make sure you have some extra room where the outlet will be installed. Last, but certainly not least, you will need the outlet itself.

Wiring an Outlet

If you are working on a wall that contains studs, begin by determining where the outlet should be located and anchor the plastic outlet box with screws to an available wall stud. When dealing with a wall made of cob, straw, cordwood, or other alternative materials, however, wall studs are often non-existent, thus knowing where your outlet boxes will be located ahead of time is important. Wherever you plan an outlet, make sure that there is a place to anchor it. In straw walls, this anchor is often a wooden stake pounded into a bale before the finish is applied; with many other alternative materials, you need to plan even further ahead and set the box into the material as the wall is constructed.

The box will have two removable plastic tabs on its top: the cable will span between the outlet, and the breaker box will enter one of these holes. Knock out the plastic tab and pull the cable through so you have about four usable inches. Using a utility knife, carefully cut the casing on the cable to reveal three smaller wires: one hot, one neutral, and one ground.

Your box is in place, and your 12-gauge cable is in the box and cut to reveal the three wires inside. Now, you are ready to reveal the copper inside each wire. Use your wire stripper to remove 1.25 inches of the insulation on the hot and neutral wires. Note that if you cannot eyeball 1.25 inches, most outlets will conveniently have a labeled gauge on their back side of exactly this length. Once the copper inside is exposed, simply wrap that copper around the appropriate connecting screw and tighten the screw by hand until the wire is securely in place.

It is important to observe that many newer outlets have push-in connectors in addition to or instead of screws to wrap the wires around. For these outlets, remove 0.5 inches of the casing and insert the exposed bare wire into the appropriately labeled hole. Many find that push-in connectors are easier to use than connecting screws.

The outlet you just wired is fully functional, but it also ends the circuit. In most cases, you will want to put more than one outlet on a circuit to be able to plug in more than two devices per room; thus, the previously described wiring technique needs to be amended a little so that the circuit can include multiple outlets. The last outlet on the circuit will be wired as described before, but for outlets in the middle of a circuit, there are a couple of extra steps.

The outlet at the end of the circuit has only one cable coming into it; for an outlet in the middle of a circuit, there will be both a cable that comes into the outlet from the hole in the top of the box and another that goes away from the outlet through the bottom of the box. In most cases, this means that you will have three wires coming in, and three wires going out. So where do these extra wires go?

Notice that your outlet has not one, but two hot wire attachment screws. You will also have two neutral wire attachment screws. If looking straight at a vertical outlet, you will attach the hot and neutral wires that come in through the top of the box to the higher corresponding screws, and the outgoing hot and neutral wires that will leave the box through the bottom to the lower corresponding screws.

If there is a tricky part to this, it is dealing with the ground wires. Unlike the hot and neutral wires that all have their own screws, you have one incoming ground wire and one outgoing ground wire, but only one corresponding ground wire screw. This is where the practice of pig tailing comes in. Pig tailing is the process of taking two or more bare wire ends and creating a functional connection between them that a current can pass through using a wire nut; in essence, pig tailing is splicing using a wire nut. So what is a wire nut? It is a small, conical piece of ceramic or plastic that accepts multiple bare wires and twists them together to form a connection. To keep your ground wire functioning properly, a bare piece of wire is attached to the ground screw on your outlet; this bare wire is then put into a wire nut along with both the incoming and outgoing ground wires. The wire nut connects the three wires together and keeps your system properly grounded.

With this understanding where the wires should go and what they should do, let us now discuss the extras that make situations unique. For example, in addition to the screws on the side of your receptacle, there may also be spaces on the back of it to accept the hot and neutral wires. These can be used instead of the screws on the side, and some people find that they are easier. If you are using metal boxes instead of plastic ones, you can pigtail yet another ground wire in your box to increase safety; in some cases, building code may require you to both use metal boxes and to use this additional grounding method.

Those outlet boxes can be cramped spaces, but it is important to keep the wires inside them as organized as possible. Always make sure that your bare ground wires are kept away from the hot terminals.

Wiring an Overhead Light and Wall Switch

Wall switches and overhead lights have in the past been wired with 14-gauge cable. Today, however, many homeowners and building codes prefer to wire all normal outlets, switches, and overhead lights with 12-gauge wire. This will make buying cable in bulk easier, and it may even increase safety and performance. There will be a slight price difference, but this may be negligible, depending on how much 12-gauge wire you can purchase at once.

The following wiring description is for a two-way switch, which means that a single overhead light will be operated by a single wall switch. Three-way and four-way switches are more complicated and are typically not a necessity, though building codes may occasionally require them. The most common place you will find a three-way switch is in a stairwell where a single light can be turned on and off from both the bottom and top of the stairs. Excellent descriptions of the several options for wiring a three-way — and other useful material — can be found online at The Home Improvement Web Directory, **www.homeimprovementweb.com.**

To begin, you will need a wall switch and a light fixture. From the breaker box, you will run a length of cable to the switch and connect the black "hot" wire to the lower screw of the switch.

Next, you will run a length of cable from the switch to the light fixture, connecting the black hot wire both to the top screw of the switch and the corresponding hot receptacle of the light fixture. When the switch is in the "on" position, electricity is allowed to flow from the breaker, through the switch and to the light; in the "off" position, the flow of electricity is broken in the switch and the light stops receiving electricity.

The neutral wire does not connect to the switch at all. After being connected to the proper receptacle at the fixture, the incoming and outgoing neutral wires are simply connected together at the switch using a wire nut. The ground wire will be connected at the ground screw in the switch and at the ground receptacle in the fixture; for convenience, a small length of bare wire is connected to the switch's ground screw, then connected to both the incoming and outgoing grounds using a wire nut.

Running and Mounting Cables

There are two basic options for running cable from place to place. The first is to run cable inside of the walls, which is how it is done in traditional, residential construction. In stud and drywall walls, as well as in most cordwood construction, there is a cavity inside the walls where cable can be run. For structures made of earth bag, straw bale, and cob, no cavity exists, so the cable is simply laid between courses (as in straw bale) or imbedded in the material (as in cob). Sometimes code will require the cable run inside of alternatively constructed walls to be UF (direct burial) cable; such cable is designed to be buried underground so its sheathing is particularly durable and resistant.

The paths of cables inside any alternatively constructed wall, as well as the position of any outlets or switches, should be planned before construction begins. While traditional drywall and stud walls can have a wide window where cable can be run, it is difficult to start installing wiring inside a straw, cob, or cordwood wall once you have built it above the height where the cable should be located. After a non-traditional wall is complete, it is difficult – sometimes near impossible – to go back and install wiring inside, so even if you are hoping to "rough it" for a while, it is wise to install cable in case you do ever decide to run electricity through your home.

The second option for running cable is to mount it on the surface of a wall. This has few benefits for the traditional home builder but can be quite convenient in an alternative structure. For example, rammed-earth walls need to be tramped and beaten to be stable, and any wire running inside such walls will be abused in the process. If you need concealed wiring, you can use heavy masonry-grade conduit to guard the cables in rammed-earth walls, but if you do not mind a little character, surface-mounted wiring may make the process easier. Installing wiring on the outside of a wall allows you more leeway during the construction process; it may be the only reasonable option for retrofitting wiring in some alternative structures, as well.

Installing surface-mounted wiring does not mean that you must have visible cables running all over the place. You can easily install all your wiring using the exposed-wire molds you often see in commercial and industrial settings. The molding around the wire is an additional material and means that the total cost of your electrical system will increase, but the ease of installation, simplicity of alteration, and overall aesthetic appeal of such moldings makes them an option worth considering. You can make your own wire molding or conceal wiring inside PVC, which can be attached to the wall with brackets and painted, if desired, but when ready-made wire molding is available — and attractive — one could make a strong case for going ahead and purchasing it.

If the connecting of outlets and switches still makes you uncomfortable, you can save money without dealing with a single hot wire by putting your lengths of cable or conduit in place before the electric company is even contacted. This will take some effort, but the upside is that it will save the electrician you hire some time. Running lengths of cable can be a tedious, long process; if you have it done before your electrician gets there, he or she can immediately start making connections and installing outlets and switches, which will save you money on the total bill.

Alternative Sources for Electricity

The owner-builder will have far more options for his or her home than anyone purchasing an existing house. But while this variety of choice presents itself in many ways, one of the most compelling is the option to choose where your energy comes from.

It should first be made clear, however, that at this time, connecting to an electrical utility is generally less expensive than producing your own power — at least in the short term. One exception can be found in very secluded areas where the initial investment of installing your own electricity producing system might be close to the cost of having a power company run a new line to your property. As prices of alternative energy components become more manageable and the number of federal and state government incentives increases, the added costs of powering your home with alternative energy will decrease, too. Currently, most folks who opt for alternative energy do so for environmental reasons rather than financial ones, but in some cases, wind, hydroelectric, and solar systems can pay for themselves over time. This becomes even more likely when you eliminate the cost of professional installation.

Electricity can be generated from different sources, some greener than others. Some areas are rich enough in solar, wind, or hydroelectric potential that no other energy source is necessary. In many cases, a combination of two fairly reliable, natural energy sources will be able to provide adequate power for a family who is conscious about their electricity consumption. In yet another scenario, many homeowners have found that installing the components needed to harness a single alternative energy source might not get rid of all their electricity needs, but it will reduce utility bills enough to make the investment a feasible – and even profitable – one.

Hybrid systems incorporate a blend of energy sources, such as solar photovoltaic (PV) panels, wind generators, or hydroelectric, depending on the location and availability of sun, wind, and water. Back-up propane is often used to carry additional loads or in certain seasons, such as winter, when heating the home is required. The type and number of appliances and lighting will determine how much electricity a home will need. Using small, energy-efficient clothes washers, refrigerator/freezers, and light bulbs will make energy needs easier to fulfill.

Solar Power

Advances in solar energy technology for the home have made it much more accessible in recent years. The technology for solar electricity was actually invented in the 1870s, but it was not used much until the 1950s. Solar PV panels gained recognition as a means for providing electricity for the home in the 1970s and 1980s. Today, PV panels are considered one preferable option for those who want to live completely off the grid, and even for those who remain on-grid but want to reduce consumption and send excess to the energy grid. Homes often do best with a combination of renewable sources of energy, especially in climates that experience periods with heavy cloud cover.

Older solar set-ups and newer set-ups in very remote locations depend on batteries to store the electricity produced by the solar panels. Essentially, sunlight hits the panels; the panels convert the sunlight to electricity; and the electricity is pumped into a large set of batteries that can be used to power various electrical appliances. Today, solar panels are often connected to a power grid instead of a set of batteries. The electricity they produce goes to an inverter, which converts it from a direct current (DC) to an alternating current (AC) that makes it suitable for household appliances. When no electricity is used, power produced by the panels runs to a utility company, where it can be used to power someone else's house.

Grid-connected systems provide you power when you need it, and the power company pays you for the excess electricity when you do not; none of the electricity your panels produce is wasted. Because solar panels do not produce electricity at night, and their production decreases significantly on cloudy days, you will still use some grid power; depending on your home's energy consumption, however, your utility bill might be a small fraction of what you used to pay. One thing to note: Grid-connected solar systems must be professionally connected. Also, some government incentives associated with solar panel installation are only available when a certified professional installs the panels, even if they are connected to a grid.

One of the main factors to consider when installing solar panels, a wind turbine, or hydroelectric system is that there is no way to tell how much a system will cost without a thorough examination of the property it is to be installed on. Though some companies and organizations will give you a number for the average cost of installing such a system, your particular situation will be unique, and you must self-determine whether the process is worth it to you.

There are several items you need to examine on your property when considering solar panels as a primary or supplementary energy source. The amount of sun is, of course, the biggest factor. Maps that display the average amount of sunlight hours in a given area of the country are available online. In the U.S., check out **www.solarcraft.net/sun-hours-map.htm**. These can give you a general idea of where the solar hot spots are. This map will not mean a thing, however, if your house is oriented to the wrong direction or is shaded by trees for most of the day. Planning to install solar panels is a large and delicate process, and the cost of having a professional do it right should be included in your overall estimate for the installation.

While grid-connected solar panels absolutely must be connected by a professional, off-grid systems— and many non-connecting steps of on-grid installations – can be done on your own because they are entirely separate from utility lines. If you want to install your own solar panels, there are a few aspects to consider. The decision to install solar panels can be made well after your house is complete, but there are several advantages of making your mind up during the designing phase of construction. Solar panels that are installed after the house has been built will simply have to be fit in wherever possible, which is usually on the roof. This frequently leads to a solar design that is passable, but not optimal. By including solar panels in your initial design, you will have the chance to find out where they will be the least intrusive, where they will get the most sun exposure, and even where they will look the best. By planning early, you can avoid the expense and hassle of retrofitting.

In the Northern Hemisphere, solar panels generally perform best on a home's south-facing side, where they will be exposed to the most sun. Of course, obstacles like chimneys, trees, and other landscape features might block sunlight in some situations; this, again, can often be avoided by planning panel placement in the early stages of design.

Once you have become comfortable handling tools like power drills and stud finders, and you have a better idea of how your home's wiring works, installing solar panels should not be as difficult. You do have to pay special attention to the angle your panels are installed at, so brush up on your math skills. You also need to be careful when installing panels on a completed roof, as the holes created by the posts your solar panels will be fixed to can cause leaks if they are not properly sealed.

Hydroelectric

Hydroelectric energy works by using falling or rapidly running water in a river or stream that is concentrated in a specific place, such as a dam, to turn a turbine. The turbine then drives an electric generator. Small hydroelectric plants can produce up to ten megawatts and can be connected to the grid or be off-grid, and produce electricity for one home or for a community. It is possible, in some cases, to redevelop a former waterwheel site for electrical power.

Hydroelectric power produces no pollution, is renewable, and can be a constant source of power as long as the water is plentiful. However, water rights are tricky, and because people are protective of their property, ensure that a potential hydroelectric site is available. It may even be worth getting legal advice before getting started on a project to determine the water rights of the area. One environmental drawback to hydropower is the degree of impact it can have on the nearby ecosystem because of the amount of infrastructure required to set it up.

Wind

Wind turbines are changing our landscape — and in many communities, changing the way our electricity is produced — but understanding just how effective they are can be confusing. The fact is, however, that wind turbines are growing in popularity and are an important piece to the puzzle that is created by modern energy needs.

Wind is used to generate power for about 1 percent of the world's total electricity use. It is used for about 6 percent of electricity use in Germany and about 19 percent in Denmark, according to the Global Wind and Energy Council. Generators use large blades to turn turbines, which then rotate magnets to create electricity. While a fairly steady supply of wind is necessary to make a turbine worthwhile, an average speed of about 10 mph is all that is required. If the turbine is not required for a home's total energy needs, then intermittent wind is usually not a problem.

How much power does a wind turbine produce? The short answer to this question: plenty. Though statistics vary, many would put the amount of electricity the average American household uses somewhere in the 10,000 kilowatt-hours a year range. According to the American Wind Energy Association, the largest wind turbines are each capable of producing enough energy in a year to power more than 1,400 average households. These large turbines, however, are hardly suitable for residential installation. Residential turbines are smaller and produce less energy than their larger relatives, but they are also far less expensive and, in the right situation, they can handle a significant percentage of a home's yearly electrical needs.

We can control a lot of factors in our home creation, but the weather is not one of them. For many owner-builders looking to use wind power, the biggest question is this: What happens when the wind stops blowing? Of course, the wind is not always going to be strong enough to allow a wind turbine to produce to its full potential; that does not mean, however, that the lights go out when the turbine stops spinning. Most residential wind turbines work in conjunction with the local utility company. When the turbine is producing more than enough energy to power your home, the extra energy is automatically sold to the utility company. When the turbine produces too little, your home switches over to power the utility company produces. This relationship ensures that your home always has the electricity it needs, and it can significantly lower your energy bill.

According to the American Wind Energy Association, "A small turbine can cost anywhere from $6,000 to $22,000 installed ... The wind system will usually recoup its investment through utility savings within six to 15 years." It should also be noted that handy homeowners could opt to purchase their turbine directly from the manufacturer and put it up themselves to further reduce costs.

Not every home in America is going to fit the bill when it comes to residential wind power. Though your current electrical wiring is probably not going to pose a problem when it comes to installing wind turbines — they can be added to most home wiring systems without making any major changes — the location of your house may be a deal breaker. If your home meets the following criteria, however, purchasing a wind turbine to supplement your electrical needs may be worth looking into:

1. Your home is not in a city or crowded suburb.
2. Your home sits on at least an acre of land.
3. The area you live in has at least a 10 mph-average wind speed.

While these are not the only factors that will come into play when installing a wind turbine, they are a good start for anyone wishing to reduce pollution; wind turbines, like solar panels, produce electricity without producing emissions. Energy bills decrease at the same time, and if current trends continue, there is a good possibility that wind power will become even more popular, more accessible, and less expensive in the future.

Turbines used to power homes usually have a capacity of about 100 kW or less. Their power can be stored for power outages, and turbines that are connected to the grid can use grid energy storage to displace purchased energy when the turbine is running. PV panels or even batteries are used to offset periods when wind is less available. Some people even use smaller wind turbines to charge a small battery that can be used as needed. The costs for using wind power are primarily up-front in construction and materials; once a system is in place, wind energy is an inexpensive method of producing electricity.

Heating and Cooling

Most modern homeowners in the United States are adamant advocates of the temperature-controlled living space. When it gets hot, we cool ourselves down with air-conditioners; when it gets cold, we warm ourselves up with furnaces, boilers, and electric baseboards. However, conventionally built houses that are not specifically designed for energy efficiency are often incapable of holding that heated or cooled air indoors for long periods of time, which leads to uncomfortable living spaces, higher energy bills, and a lot more energy consumption than necessary!

In the alternative building world, special attention is paid not just to where materials come from and how they are used, but how comfortable they keep the temperature inside of your home, too. There is certainly no harm in installing a cooling system in an alternatively constructed house, and having a heating system in place is an absolute must in many areas; however, if your home is designed and constructed properly, there is a good chance you will have to use these systems less frequently than you would imagine.

Due to the high efficiency of many alternatively constructed homes, you will often find that the furnaces, boilers, and air conditioning systems seen as selling points in a conventional house are replaced with stoves, swamp coolers, and other contraptions that do the job just as well and use a fraction of the energy. This is excellent news for owner-builders, as these devices are often easier to install than their conventional counterparts.

CASE STUDY: OUT WITH THE OIL, IN WITH THE NEW

Pat Murphy, Executive Director
Arthur Morgan Institute for Community Solutions
www.communitysolution.org
pmurphy@communitysolution.org
P.O. Box 243
Yellow Springs, Ohio 45387

Five years ago, Pat Murphy learned about the threats of continuing dependence on fossil fuels, namely the coming global oil production peak and catastrophic global climate change. He became the executive director of a small nonprofit called Community Solutions, which promotes the re-emergence of the small community and a more agrarian, low-energy way of life to address peak oil and climate change.

Community Solutions educates, advocates, and researches ways to dramatically reduce energy consumption. It provides knowledge and practices to support low-energy lifestyles, with a primary focus on reducing energy consumption in the household sectors of food, transportation, and housing. Community Solutions designs and develops solutions to the current unsustainable, fossil fuel-based, and overly consumptive way of living and seeks to raise awareness of the coming peak and decline in global oil production and the threat of global climate change.

For housing, Murphy's interest is in the Passive House building standards and deep energy retrofits for existing buildings. The Passive House, which was developed in Germany, reduces a building's heating and cooling energy use by up to 90 percent. Compared to programs like Energy Star and LEED certification, which only save, on average, 15 to 30 percent of the energy used, the Passive House actually cuts energy use by the amount required to curb carbon dioxide emissions.

"A Passive House is a very well-insulated, virtually air-tight building that is primarily heated by passive solar gain and by internal gains from people [and] electrical equipment," according to the Passive House Institute U.S. With so few energy losses, Passive Houses can be heated with an extremely small external source — or none at all.

CASE STUDY: OUT WITH THE OIL, IN WITH THE NEW

Changes in the green building field include a huge increase in the number of articles and marketing information for new green building techniques and products. Yet there have been few changes in standards, codes, or practices. But in general, interest is increasing along with higher fuel prices as homeowners begin to see the economic benefits of making their homes less energy consumptive. As far as drawbacks, the standards for green building are too low. Much of the movement is based upon "feel good" ideas with little energy savings. A much stronger standard beyond "green building" is needed to avoid disastrous climate tipping points. Because about 40 percent of total U.S. energy consumption is used to operate the more than 100 million existing buildings, major energy retrofitting will be needed to reduce our energy dependence.

Standard Heating and Cooling Options

For most conventionally constructed homes in the U.S., a heating system generally means a furnace and forced-air set up. Here, the furnace burns oil or gas to create hot air; this air is then pushed around the house through a series of ducts, leaving occupants warm and cozy. The problem with forced-air systems is that they are expensive and complicated to install and require the burning of fossil fuel. Setting up a forced-air system is a large job, and the components are expensive; add that to the fact that burning fossil fuel for heat is hardly a sustainable practice, and it is easy to see why so many owner-builders opt for a different route.

When you look at standard cooling techniques in conventional building, you run into some of the same problems you do with conventional heating techniques. A condenser located outside the house often powers central air conditioning. This device uses massive amounts of electricity to change the temperature of the air and pumps it throughout the home's living areas via a bulky system of ductwork. Running on the same principals as a refrigerator, these condensers are responsible for a great amount of energy consumption – and some rather hefty electrical bills. In most cases, a professional sets up conventional heating and cooling systems. The ductwork for these systems can last a while, but the condensers and furnaces that power the systems will require repair, regular maintenance, and eventually replacement. Some recommend replacement of furnaces and condensers after as few as 15 years.

Passive Solar Heating

Passive solar heating is a design consideration rather than a matter of buying the proper piece of equipment. While other forms of heating a home can typically be added after a structure is built, passive solar heating must be considered at a structure's conception.

This method of heating focuses on bringing heat from the sun indoors and keeping it there. The first step in such designs is including a large number of windows – often, one entire side of a structure will be made of glass. Flooring and high-thermal walls then absorb the heat let in by these windows, and the house will retain heat for long periods of time, releasing it slowly.

For owner-builders in colder climates, passive solar heating can make a large difference in the amount of heat that must be created by electricity or burning fuel. In warmer climates, passive solar heating is likely to eliminate any need for supplemental heat sources.

Many alternative builders prefer passive solar design for good reason. Not only does this technique reduce heating costs by harnessing solar energy, but it also reduces the energy loss inherent in thermostat-controlled systems. If a central heating system and thermostat exclusively control the heat inside a house, the system is constantly turning on and off while trying to keep temperatures comfortable. By using passive solar design, however, your home will not just have higher amounts of natural heat; it will have a more consistent temperature, as fluctuations between cold and hot will be less dramatic.

To be effective, passive solar design involves optimal window placement and incline, a conscious layout of the house so that the heat collected from the window wall can be easily distributed, and good planning of walls and flooring so that heat is retained where desired and reflected back into the living space where necessary. Complex as it is, passive solar design is the most eco-friendly heating option available, and if planned well ahead of construction, it will cost little (or no) more than any other design. Creating a passive solar design on your own can be done with a great deal of research, but there are professionals out there who deal specifically with this discipline and know the best floor plans, the best materials, and the most important considerations to make the design successful. Consulting with, befriending, or even hiring one of these professionals may be the wisest choice.

Geothermal Heat Pumps

Geothermal heat pumps are an efficient, effective, and eco-friendly way to both heat and cool your house. Unfortunately, they also require professional installation – and cost more than other options. The pump unit itself is not extremely costly; the real expense is in the loop, or the deep hole in the ground that is needed for these systems to function. Like many other alternative energy options, if a geothermal heat pump is used frequently enough for long enough, it can conceivably pay for itself over time; however, as these systems can sometimes be extremely expensive to install, it is worth the time to get professional installation estimates, calculate your potential utility bill savings, and compare how much of a return you will gain on this investment.

Once a geothermal system is in place, it can last a long time. Many companies claim that the loop will last a lifetime, and the actual heat pump unit can last for decades. Both geothermal and conventional heat pumps work not by heating or cooling the air, but by moving the desired temperature air from another area to the inside of the house. This requires electricity. A geothermal heat pump will not be a 100-percent natural form of heating and cooling due to the electricity it uses, but when compared to conventional furnace or condenser systems, the temperature control you get from a geothermal system is far greener.

Wood, Coal, and Biomass Heating

One of the cheapest and easiest-to-install options for heating your owner-built home is a stove. When properly situated in your floor plan, it is often possible to heat your entire house with a single stove, though very large homes — which tend to be a rarity in the alternative and owner-built world — might find a second heating source that better fits the situation. When set up by a professional, some stoves can provide your home with hot water in addition to heating the indoor air. In many cases, owner-builders have found that the cost of stove heating is significantly lower than that of conventional heating, both in installation and in yearly fuel costs.

If you plan to install your own stove, the biggest factor in how efficiently it heats your home will be its placement. A central location on the first floor of the house is generally accepted as the most efficient place to put your stove, but many have made do with stoves that are not directly positioned in the home's center. If you plan on having a professional stove heating system installed, talk to your intended installer early because the system might involve certain considerations in your home's design to work efficiently. These professionally installed systems work well and do not cost much but require ductwork and blowers, much like conventional forced air systems, to operate at full capacity.

Wood is the oldest heating material in human history, and the fact that many alternative homes are situated on land where wood is plentiful means that, with a little labor and foresight, you can have an efficient and inexpensive heating source without depending on outside fuel. Your garden-variety wood stove depends on heat radiated from within the stove to heat the house, so the farther away a room is from the stove, the less heat it will receive. This effect is significantly reduced in round houses where the stove is placed in the direct center of the floor plan, but if the stove is large enough, burns hot enough, and your walls and roof are insulated enough, the amount of cold spots you will find around the house will be surprisingly few, even in a rectangular home. All that is needed to install a wood stove is a stove and a chimney for smoke to escape.

Coal stoves burn extremely hot, and because coal is an inexpensive fuel source in many areas of the country, those in very cold climates will find that their coal stove is less expensive to operate on a yearly basis than more conventional forms of heating like gas, oil, or electric. Coal has not historically been known as the most eco-friendly substance in the world, but today's coal stoves burn hotter and cleaner than their predecessors. And the use of anthracite coal rather than softer varieties of coal leads not just to a hotter fire, but a cleaner one, too. Anthracite coal has a low — generally about 1 percent — sulfur content and has been said to have a negligible effect on the environment. Coal produces hard ash called cinder, and those who burn coal as a fuel source will have to somehow dispose of this by product. In snowy areas, it can be used to provide traction on icy driveways and walks, but you will probably have a lot more cinder than you will ever have use for if it is your primary heat source. In many cases, cinder is simply put on an out-of-the-way pile and forgotten about.

Biomass fuel is generally formed of one or more kinds of plant material. It is a relatively new trend in heating fuel but is quickly gaining popularity in North America for several reasons. When compared to a traditional wood-burning stove, biomass stoves produce more heat while creating fewer pollutants. Some biomass stoves are similar in operation to traditional wood or coal stoves, but others are more complex. To add to its appeal, biomass fuel comes from refuse like sawdust that is created in other processes; when compared to cutting down trees or mining coal, the creation of biomass fuel looks like a friendly practice. If you are looking into stove heating for your owner-built home, there is no harm and plenty of benefits in doing price and performance comparisons between biomass stoves and wood or coal burners.

Roofing

A house is simply not a house without a roof. Roofing can protect you from rain, sun, wind, and cold, and the area in which you live will determine your roof's most important function. In dry, arid areas, your roof should primarily focus on providing shade and keeping you cool; in areas where rain is a pain, your roof's most important function will be keeping you and your belongings dry. In areas like New England, a roof's biggest job is keeping you warm. Because your roof's functionality changes by location, it is important to consider the benefits and drawbacks of the materials you use on your roof and the way it is constructed – not just because of your budget or tastes, but also in respect to your climate and weather patterns.

Your roof construction is not always the final step in the building process. In fact, some scenarios lend themselves well to installing the roof even before you build the exterior walls. When and how you erect your roof will depend on the type of structure you are building. For example, the massive walls of a cordwood house constructed with built-up corners are meant to support the weight of the roof, so the roof goes on after the walls are completed. On the other hand, timber-framed constructions with cordwood, straw bale, or cob walls that are not meant to support the weight of a roof can often benefit quite a bit from putting up the roof early in the project. This will go a long way in protecting the elements of the house underneath it from Mother Nature during the long building process.

Deciding on the Pitch of Your Roof

The area where you live, the roof's material, and its function will determine the best pitch (steepness) of your roof. In many cases, the steeper a pitch, the more difficult the roof is to construct. In areas where rainfall and snow are scarce, you can easily get away with a roof that has very little pitch. In a colder, wetter area, a steeper pitch will allow precipitation to slide off, which will reduce moisture issues and minimize the added weight of a heavy snow. Fewer issues and less strain can help extend the life of your roof and sometimes the materials under it.

The pitch of a roof is identified by a set of two numbers. The first number is the vertical rise, and the second number – typically 12 – is the horizontal length. The flatter the roof, the lower the first number. A completely flat roof would rise 0 inches every 12 inches; an almost flat roof may rise 1 inch every 12 inches (1-12); and a steep roof may rise 12 inches every 12 inches (12-12). A roof in an area that receives heavy snow should have a significant pitch to decrease the weight the roof will carry after a big storm; about 6-12 would be an acceptable pitch in this kind of environment. A 2-12 roof will require less material, time, and money to erect than a 12-12 roof, but a steep, 12-12 roof will shed precipitation more efficiently.

Some roofing materials perform best at a certain slope. Living roofs, for example, should be constructed on a very low slope to ensure moisture retention and to keep gravity from pulling the plants off. On the other hand, a metal roof can be built on almost any slope. This is one of the many reasons that alternative builders often choose to use this material.

Picking Your Roof Material

Some roofing materials are appropriate in just about any environment. Asphalt shingles, for example, tend to perform pretty well in climates from wet to dry and from hot to frigid, which is why they are the most widely used roofing material in the U.S. However, when it comes to alternative home building, asphalt is generally not the preferred material among owner-builders. It is less eco-friendly than many other options, and it is also likely to wear out faster. For most owner-builders, putting up one roof in a lifetime is more than enough. This section will describe which options alternative builders tend to flock to and give you good reasons why.

For alternative owner-builders in the U.S., the options are many, but the ultimate choice often comes down to only two prospective materials: living (roofing designed to be covered in plants and earth) and metal. Metal roofing is lightweight, easy to work with, appropriate for just about any environment, and can often last 50 years or more. Its durability reduces the amount of roofing material your home will require over its lifetime, which leads to energy savings in commercial material production, and you may also be able to find metal roofing products made with recycled content. Living roofs are far more difficult to construct, but when done properly, they never need to be replaced, will hold up under all kinds of conditions, and will provide unmatched insulation while remaining a sustainable, eco-friendly material. When compared to the relatively short lifespan of asphalt, the expense and hassle of slate and tile, and the maintenance required for wood, metal and living roofs are the clear choices for owner-builders who want performance, longevity, and affordability.

Attaching Your Metal or Asphalt Roof

While the framework of your roof can vary in size and scope, the ideas involved in constructing a functional roof are fairly universal. Your roof needs to be firmly anchored so that a strong wind will not carry it away. Its weight should ultimately be borne by your foundation and the earth underneath your foundation so that it will not shift. Its surface should be sturdy enough for you to walk on if repairs or additions are necessary. Finally, your roof should be constructed so that water cannot easily infiltrate its surface and harm the other elements of the house it protects.

Creating a sturdy roof can be accomplished in several ways but the most effective involve a hefty number of triangles. Triangles are sturdy shapes, and a good roof will have triangles in all the right places. In stick-framed houses, you will be constructing a long series of many lightweight triangles to create your roof. In post and beam houses, you can often get away with having fewer, though much heavier duty, triangles. In a roof, the bottom side of your triangles will span the width of your house; these are called collar ties. The two, generally equal, sides of your roofing triangles that meet in the center and form the pitch of the roof are called rafters. When the two rafters are joined to each other and are both fixed to the collar tie, they form the most integral triangles in your roof: the trusses.

Trusses can be constructed in place, but when creating trusses from lightweight components like 2x4s and 2x6s, it is often easier to keep your measurements – and therefore the dimensions of the trusses – precise and accurate if you build them on the ground and then lift them up. This technique is less feasible when the trusses are very heavy; for instance, if you are creating timber trusses from whole logs, it would be difficult to lift a completed truss up to where it needs to go and much easier to build them in place.

Making sure that your roof is sufficiently supported is done in one of two ways: It is either anchored to the load-bearing exterior walls of your house, or it is anchored to your home's skeletal framework. In both cases, your foundation and the earth below it ultimately carries the weight of the roof.

If your walls are load-bearing, the easiest way to firmly fix a roof to them is to create a top plate. A top plate is basically a series of sturdy pieces of wood that are fixed on top of your home's load-bearing exterior walls. The wood top plate allows you to use anchoring hardware to attach your rafters and collar ties. It is easier and more effective to attach a wooden truss to another piece of wood than it is to fix that truss to a mud, plaster, or straw bale wall.

Homes without load-bearing exterior walls must depend on the framework to keep the roof from blowing away (or the shell of the house from crumbling under the weight of the roof). Whether timber, stick, or post-and-beam framing is used, the weight of the roof in a framed home rests on the framing components, so they must be sturdy enough to support its weight. The vertical members of your framework that are on the perimeter of the house will be the ones that are primarily responsible for bearing the roof load; these vertical members will be connected to each other in two ways: Each vertical member will be connected to their adjacent vertical members by a top plate and to the vertical members directly opposite them by a cross tie.

Whether you have load-bearing or non-load bearing walls, or you have stick, post and beam, or timber framing, a ridge beam will be installed at the apex of your roof. At the center where the two slopes of your roof meet, the ridge beam provides stability and ties all of the trusses together. In many cases, notches are created both at the apex of the trusses and at corresponding points in the ridge beam so that they actually lock in place in addition to being joined by hardware lice nails and screws.

Once the ridge beam is installed, you are ready to install the roof decking. The decking is the material that provides a continuous base for your roofing material to be installed on. Decking is generally made of plywood because of its strength and durability; with plywood decking underneath and a sufficient number of trusses supporting it, you can rest assured that walking on your roof can be done without fear of falling through. When decking is properly installed, you will finally have a roof over your head that can protect materials and supplies underneath it from precipitation. It is a good step to accomplish because it really feels like progress has been made after the decking goes up, and if you have been storing materials in another area to keep them dry, you will now have a faster, less cumbersome cleanup ritual. It should be noted here that metal roofing could be installed over "stringers" (1x2s) that are nailed or screwed to the rafters. This option uses less wood and costs less money, but the resulting roof is weaker and could be damaged if you walk on it.

Once the decking is installed, you will be able to install your roofing material. Asphalt shingles are typically nailed or stapled in place in staggered rows, starting at the edge and ending at the ridge so that each row of shingles is overlapped by the one above it. Metal roofing is typically screwed rather than nailed in place; it is very important to make sure that each screw in your metal roofing is made water-tight, either with a washer or using a material such as tar to fill in any gaps that might occur. The installation of particular brands of roofing materials – be they asphalt, metal, or other – should be done by any applicable manufacturer's guidelines. Unfortunately, the warranty of many roofing systems is voided when not installed by a professional roofer.

Living Roofs

Living roofs have many benefits. They are excellent insulators, have an interesting and natural appearance and, if properly installed, will last a lifetime and require little to no maintenance. Living roofs resist heat transfer because the plants – which also help protect the inside of the home from rapid temperature fluctuations – absorb heat and provide evaporative cooling. The plants and thick layers of growing medium protect the roof from wind and sun, which are two typically destructive forces for roofs.

Preparing the structure for a living roof can be tricky, as a roof with plants growing on it needs to maintain a certain level of water to keep the plants alive. They absorb water from the atmosphere, requiring the underlying structure to be well-supported. A living roof is far heavier than any conventional roof, so planning your support structure is extremely important and should be done early. If you are planning on installing a living roof, make sure a qualified contractor takes a look at your plans and does the math. All that weight will need to be carried for as long as the house is standing, which means that the sticks, posts, or load-bearing walls need to be up to the challenge. What might look like a sturdy support system to the owner-builder may have some obvious flaws that a professional can spot immediately. A lot of time and effort will be needed to construct a living roof, but if your structure is not able to support the weight, that time and effort will be for nothing.

After you have discussed your support structure with a knowledgeable professional, work on finding materials for your living roof. Several designs of living roofs exist, but just about all will contain a few key materials. From the bottom up, you will need at least:

1. Decking (often plywood or tongue and grooved planks) to create a strong, even surface. The decking is installed in the same manner as it would be for a metal or asphalt roof, but because of the heavy weight it will bear, it requires significantly more structural support underneath.

2. A waterproof membrane – often a pond liner, Bituthene, or EPDM rubber material designed just for this purpose – to protect the decking from rot and decay. The membrane needs to be kept water-tight, so it is often fixed to the decking with a mastic sealant. Make sure that this membrane is not applied on a surface where nails or screws stick out. If the membrane is pierced, the resulting leakage could rot your decking.

3. A layer of insulation – typically of the rigid board variety – will then be installed to enhance the efficiency of the system.

4. A root barrier and fabric filter is added to keep roots and loose soil from hindering drainage.

5. A layer of soil of varying depth and composition is then added. This soil layer is heavily dependent on the area in which you live and the type of plants you plan on having on your roof.

6. Appropriate vegetation is then added, watered, and grown on top of the roof. Some systems require less watering than others; a few might require a gardener, landscaper, and regular mowing.

Pitched roofs that rise more than 2 inches for every 12 inches will require a layer of binding material to keep the plants from sliding off. Try GardNet, Geoweb, and Enkamat for binding material made specifically for sloped living roof installation, though some green builders prefer using chicken wire and other more generic products as a binding material. Unfortunately, not every pitched roof will be a candidate for a living roof upgrade; if the pitch of your roof is too steep to walk on, you can rest assured that it will also be difficult to grow plants on, too. Flat living roofs – and some pitched ones, as well – require a drainage layer to remove excess water caused by heavy precipitation, and this layer may be designed to act as both a soil filter and drainage mat.

There are two types of commonly referenced living roofs: Intensive and extensive. Intensive living roofs are often installed as much for their beauty as for their practical attributes. These attractive installations can support a wide variety of grasses, flowers, and other plants but require a heavy soil layer of up to 12 inches in thickness. Not only can these roofs be walked on, but some people do their gardening on them or install park benches to turn their roofs into a hangout spot. In some intensive living roofs, shrubs and even trees have been successfully included in the design.

An extensive living roof uses very specific vegetation. Though not as pretty as having a field of wildflowers on your roof, these hardy plants require less soil to grow (sometimes as little as 3 inches, making the roof significantly lighter than intensive varieties) and need less maintenance than the flora you will find on intensive living roofs. For owner-builder, extensive living roofs are typically easier to plan, create, and maintain; you may even be able to purchase a "kit" to install a roof of this variety.

There are plenty of different styles and designs available for those who want to construct a living roof. If you go the kit route, it is likely that you will need little professional help to get the job done. For the inexperienced owner-builder, these kits come recommended for several reasons. They are designed for easy installation; they generally come complete with everything you will need for the job, other than the tools; they will have specific and detailed directions for proper installation and maintenance requirements; and the companies that manufacture them have excellent knowledge of the system that is best suited to your area. In most other cases, where a kit is not used, professional consultation – if not professional installation – is an absolute must.

Living roofs can be wonderful additions to the green home, but because you are dealing with changing, growing, hungry, thirsty materials instead of wood, metal, or asphalt, living roofs can deteriorate in different ways. No one ever heard of a metal roof getting a disease or an asphalt roof needing fertilizer, but when dealing with living roofs, these things can happen. Consulting with someone who is well-versed in this process is a necessity if you want to avoid problems with your living roof in the future.

Insulation

The building materials used in most new houses are far from good insulators when compared to the materials used in alternative construction. About 1.5 to 2.5 feet of a thick straw bale, cordwood, or rammed-earth tire wall will provide an incredible barrier between the outside air and the interior of your home, but a little vinyl siding and a bit of drywall must be supplemented pretty heavily with insulation. Some alternative materials are better than others when it comes to insulation value. Adobe homes in cold areas, for example, will have to be bolstered with just as much insulation as most conventional constructions. This is one of the main reasons adobe homes are generally built in areas where heating is less important than cooling, and materials that are better insulators are generally used in cold-area alternative construction.

Though picking the right materials for the shell of your house can go a long way in increasing its efficiency, insulation certainly has its place in most alternative constructions. In many cases, that place is easy to identify: under the roof. If you plan on having an asphalt shingle or metal roof — two of the most common and practical choices for any home — you will definitely want to look into insulating its underside. Owner-builders will find that rigid insulation boards and rolled insulation are generally the easiest (and often most cost-effective) options for beefing up their attic insulation. The blown-in and sprayed-in insulation options described below are generally most appropriate for wall cavities in conventionally constructed houses.

Rolled Insulation

Rolled insulation is the fluffy, often pink or yellow fiberglass insulation that most of us are already familiar with. In conventional construction, it is often placed in between wall studs, rafters, and floor joists. When purchased, this kind of insulation comes in rolls that you can cut yourself, or in "batts" that are pre-cut rolls. To install it, you simply stuff it in between your joists, rafters, or studs and staple it in place; it is then hidden from view by sheets of drywall.

This type of insulation is easy for a do-it-yourselfer to install, but it has been known to cause temporary skin and respiratory irritation, especially when prolonged contact with the material occurs. Long sleeves, work gloves, and even a surgical mask can be used to reduce these irritations.

Blown-In and Sprayed Insulation

Blown-in fiberglass insulation can significantly bolster your home's insulation. In the attics and wall cavities of conventional constructions, small, loose particles of fiberglass are blown in by machine to fill up the designated area. This operation is almost always done by a professional and is not a great — or even feasible — idea for owner-builders to try themselves. Another option for blown-in insulation is cellulose, which can be made from recycled paper products, making this an eco-friendly alternative to the more conventional type of blown-in insulation.

Sprayed polyurethane foam insulation takes the concept of blown-in insulation to the next level. Once sprayed in the desired area, this foam actually expands to fill in the small crevices and gaps that other insulation will miss. This is also an operation that should be done by a professional; it is also more expensive than other types of insulation, though it does provide more insulation per inch of thickness than fiberglass.

Rigid Insulation Boards

Inexpensive and easy to work with, rigid insulation boards made of plastic foam can quickly be cut to size and fit in place. These boards can be shaped to fit just about any space with only a utility knife, and their rigid construction makes them ideal for placement around foundations and under concrete where other insulation materials simply cannot go. Most rigid insulation is not meant for direct burial, however, so if this material will be coming into contact directly with earth, make sure to purchase boards that are specifically designed to do so.

Framing

Your home's framing is basically its skeletal system. Generally made of wood – though increasingly constructed of metal in commercial settings – framing is often referred to as either stick or post-and-beam framing. Stick framing is commonly seen in conventional house construction. It uses many smaller pieces of wood – usually 2x4s or 2x6s – to create the home's skeletal system. In post-and-beam framing, a smaller number of larger, stronger pieces of wood are used to create the home's skeletal system. Timber-framed construction is basically post-and-beam framing that utilizes rough, rather than hewn, logs to create the skeletal system.

In many cases, the alternative owner-builder will gravitate toward the rustic, natural appearance and function of post-and-beam or timber framing. The smaller, lightweight components of stick framing are a little easier to work with, but the production of all those sticks requires more cuts, the use of more energy, and the creation of more waste than is required in post-and-beam (particularly timber) framing. Both methods are successful when done properly, and the location of your site will help determine what is best for your situation. If you are clearing trees on your site regardless, timber framing might be a more economical choice. If you are building an adobe house in a dry area where cactus and brush are the most plentiful flora available, having timber or expensive beams trucked in will be more expensive than going to a lumber yard for 2x4s.

Homes designed with a load-bearing shell — like some varieties of cordwood and earth-rammed tires — do not particularly need a skeletal system to support a roof, as the exterior walls will do all the work. However, because most homeowners do not want their houses to contain only one room (albeit a rather large one), interior framing is typically necessary even in buildings with load-bearing shells to create partitions between one room and another. These partitions do not need to support any weight and, truly, they can be made just about any way and from any material you wish. In a structure with a load-bearing outer shell, you can make all your interior partitions out of fabric or paper if all that you require between rooms is a visual barrier. In most cases, though, creating interior walls from drywall and wood works out best in these situations. By building such partitions, you can reduce sound transfer between rooms, have a suitable place to install wiring, outlets, and fixtures, and be able to hide plumbing components.

Typically, stick-framed exterior or load-bearing walls are created using either 2x4 or 2x6 vertical components — called studs — depending on the thickness of the insulation desired. These vertical members are set at 16-inch intervals and are joined together at the top and bottom by wall plates of the same dimensions, either 2x4 or 2x6. At corners and intersections, a three-stud post (three studs that are fixed together or a single post that is the equivalent of this dimension) is employed to provide additional support. Once the desired length of wall is constructed, it is put into place and plumbed, or made vertical. The bottom wall plate is then fixed to the foundation or subfloor on the first floor and to the floor or subfloor on the second floor with screws and brackets. Each completed wall frame is then fixed to the adjoining wall frames with nails and/or screws. The frames become walls once insulation is laid inside them, drywall is screwed to their interior, and plywood is mounted to their exterior.

Unlike stick framing, post-and-beam or timber-framed walls do not have typical designs. Posts are spaced according to how much weight they can support, and this can vary dramatically depending on the type of wood used and the dimensions of the post. In some cases, only four posts are used for an entire structure; in others, multiple posts may be needed for a single wall. This variation makes it difficult to say how many posts you will need or how far apart they should be spaced. Figuring these numbers out requires some serious calculations, so it is necessary to run your intended dimensions by a professional to ensure your post-and-beam structure design is a sound one.

Interior, stick-framed walls that are not load-bearing are formed in the same manner as their load-bearing counterparts with only a few exceptions: Most of these walls will be created with 2x4s, will be covered on both sides with drywall, and will not always contain insulation inside the wall cavity. Such interior partitions are common and often used by alternative builders for convenience and to create a familiar look on the inside of the house. But they are not the only option. If you want to construct your interior walls from the same materials as the outer shell, or another alternative material, this is typically acceptable. Straw, cob, adobe, cordwood masonry, and most other materials described in this book can be used to create interior partitions and will provide a unique look along with some additional insulation and thermal mass. The factor that typically pushes alternative builders to forgo using such materials as interior partitions is convenience. These walls take time to construct, and after creating the shell of the house, many folks are pretty keen on the ease and haste with which drywall and stud partitions can be raised.

Windows and Doors

A house is not much of a house without entrances, and it is probably more akin to a hovel if there are not any windows. It will quickly become apparent in this book that many alternatively constructed homes are similar to conventional structures on the interior, but the shell of the house is a different situation. With many techniques that do not involve a rigid wood framework to be built around, and many others with walls that are more than a foot in thickness, the process of creating openings in the shell of an alternatively constructed home can be tricky for the uninitiated.

Windows

Creating a window that is set in a box is an effective and inexpensive strategy in alternatively constructed houses. The box, generally constructed of lumber and screws, is far from costly, and the windows themselves can often be found used for cheap, or even for free. When looking for used windows, it is actually preferable to opt for old windows with wooden frames, as these will be less costly and easier to work with.

A window box is basically a sturdy wooden rectangle large enough to snugly hold the window and deep enough to span the thickness of your wall. If you are building walls that are 18-inches thick, your window boxes should be 18-inches deep. If the window you are fitting in the box is 2-feet wide and 3-feet high, including its frame, the box's interior dimensions should be just a fraction of an inch larger so the frame can easily fit into it and swing out without friction but with minimal gaps between the frame and the box.

In walls where mortar or cob is used alone, or to create a bond between the other material in the wall, the window box can be more firmly set in place by using nails, screws, or lengths of wood that will be embedded in the cob or mortar while it is still wet. These anchoring devices are known as keys and will greatly reduce — and practically eliminate — the chance of your window box shifting out of place.

In the area above where the window is to be positioned, it is generally a good idea to install a lintel, a sturdy piece of wood that is longer than the width of the window. Because a window is not much more than a hole in the wall, placing a lintel above one brings back the structural strength of the wall where the hole is. Lintels carry a lot of weight — essentially, everything that is above them, including the roof and sometimes a second story — so their size is important. Some lintels must be hefty and extend well beyond the dimensions of the window they are over. For extra insurance in cob or adobe walls, reinforce the lintel itself by adding several smaller pieces of wood, called bearers, underneath it. Bearers can also be beneficial when installed beneath the top plate that the rafters and cross ties of the roof will be attached to in load-bearing shells.

Doors

Creating a suitable entrance way is a lot like putting in a window box. The two main differences are that the box for a door is typically bigger than a box for a window, and you can use the concrete foundation as the bottom of your door's box. Because a door is a pretty big hole, using a lintel and bearers above it is a good idea. It is also wise to create the door box from thick, sturdy lumber to help support the load of the wall above it and to create a strong place to attach the door itself.

Again, you can often find cheap or free old doors if you seek them in the right places. Owner-builders could even make their own doors out of lumber scraps left over from the construction of the rest of the building. However, making a door from scraps is probably not ideal for those living in a cold area, or someone who wants a traditional knob or locking system. Doors made of scraps can have a striking, rustic appearance but are often full of gaps to let cold air in. Homemade doors are also free, unique, and easy to customize, making them a perfect fit for unconventionally shaped entryways.

Papercrete and Earth Plaster

Though not a necessity in every owner-built structure, you will soon find that common plaster and earth plaster can often play an important role in many types of alternative construction. There are several recipes that can be used to create a suitable plaster, but depending on your situation, one might be more appropriate and less expensive for you.

Among alternative builders, the debate continues as to whether conventional plaster or earth plaster is a better material to coat your walls. Earth plaster is cheaper to produce than conventional plaster and is more eco-friendly. Fans of earth plaster also claim that it handles moisture better than conventional plaster, which makes it a more appropriate choice for straw and earth-based walls. It is true that earth plaster is more hygroscopic than conventional plaster, which allows the material underneath it to rid of moisture on sunny days, but conventional plaster fans claim that if applied correctly, conventional plaster will not allow moisture to reach a wall's interior in the first place. It appears that conventional plaster is more acceptable when it comes to building codes and inspectors, but those who like earth plaster are adamant that this is a material worth fighting for, and it may be in the owner-builder's best interests to convince local authorities to allow earth plastering where it is not already an accepted practice.

Both plaster and earth plaster have been used successfully in a variety of different settings, but when it comes to covering straw bales, just about every builder who is experienced in the process will tell you that earth plaster's breathability is crucial for success.

Moisture has a way of seeping into places where it should not be, and with a material like straw that is susceptible to rot and decay when wet, it is important that moisture seepage is allowed to dry out. Earth plaster's breathability allows moisture to escape; conventional plaster is more likely to trap moisture and extend the amount of time that moisture is in contact with the straw bales. While particularly important to consider when building with straw, do yourself a favor and at least consider using earth plaster when covering any material that is subject to moisture damage— which is just about everything in your home.

Though not the main ingredient in either, water is incredibly important to both conventional and earth plasters. The right amount of water to use in your mix changes from location to location and from day to day. In fact, it could probably be truthfully stated that no plaster mixtures are ever exactly the same — even on the same site. Because of this, earth plaster recipes are basically just guidelines; following them exactly does not always produce optimal results. Try a few different mixtures and test them out on a small portion of the wall to see what works best for you. Making conventional plaster is best done by purchasing a mix and adding water to it; making earth plaster seems to be best done with your feet, but can also be accomplished with a cement or mortar mixer.

Earth plaster is essentially made from clay-dirt (dirt with a high clay content), sand, and straw. To make and place earth plaster, do the following:

Outside Walls

- For plastering outer walls, start with a mix of 10 parts clay-dirt, 10 parts sand, and 1 part straw; mix by stomping while slowly adding water. Adding a small amount of borax or hydrogen peroxide to this mixture will help prevent mold. While there are no universal proportions for earth plaster, this 10-10-1 mix is a good place to start. You will have to test and retest different proportions to find the right one for you.

- Spread your plaster on a few square feet to test your mix.

- If your test patch cracks easily after it dries, add more sand (and perhaps some straw) to your next mix. If it is soft and scratches easily, there is likely too much sand in your mix; try adding more clay-dirt or another binder, such as flour paste.

- Test your plaster's durability by rubbing it vigorously once it has dried. If it does not crumble or make dust, the plaster mixture is likely durable. Spray your test section with a hose after it has dried. If the plaster disintegrates, it is back to the drawing board. Try again and alter your mix until you have a tough, water-resistant mix.

- If applying earth plaster to a smoother surface, such as cob, some builders attach materials such as chicken wire, reed matting, or burlap to the surface before the plaster is applied to help the plaster stick.

- If applying to straw bale, it might take more layers of plaster, letting each dry in between applications, to cover the straw bale shape, if such is intended.

- For plastering cob or adobe, use a plaster with high-clay content and straw to help make the plaster easy to spread with hands. Add chopped or long straw depending on the thickness needed for the wall, such as in climates where more protection is needed; the thicker the wall, the more long straw. If little cracks appear as this first coat dries, do not worry: These will help the second coat hold.

- Adding lime to the outer layer of finished plaster on an outside wall will help protect it against rain damage, though your biggest protection will come from good design: long, overhanging eaves and sufficiently high walls.

- A minimum of two layers of plaster should be applied —three are recommended for straw bales. The combined layers of plaster on exteriors should be at least two inches thick.

- Some builders boil raw linseed oil to apply to the wall as a hardening agent and to protect the wall from unloading dust. It is boiled to prevent it from turning rancid once it has been applied. This should be done with caution, though, as too much linseed oil on the outside of the wall can have the same effect as cement stucco or plaster, detracting from the wall's natural breathability.

Inside Walls

- For inside walls, sift the soil to help refine the particles for a smoother finish. Ideally, your plaster should be wet enough to spread on easily with a trowel, but not so wet that it drops off when trying to place it on the wall.

- After spreading a few square feet of plaster, allow it to set for a bit while continuing to apply plaster along the wall. After each patch has set for a short period of time, use a damp tile sponge to smooth the surface. This will give a texture for the next layer to bind to.

- If applying only one coat, use a trowel to smooth out any ridges or bumps. One coat plasters work best on cob and adobe walls that have good shape and do not require much smoothing, but are not recommended for straw bales. Be sure to sift the clay first and use chopped straw.

- For a finish plaster, clay slip works well because of its fine texture. For making earthen walls white, purchase white clay from a ceramics store. This type of clay is already sifted and will create a beautifully smooth and attractive surface. You can also make white plaster by using a mixture of white sand and lime.

While some builders add natural pigment to the last layer of plaster, earth imparts its own color to the walls, which becomes more noticeable as it dries. It might be worth waiting to see what color develops as the plaster dries before you begin worrying about adding pigment. If any mold develops on the plaster once dried, cleaning it with hydrogen peroxide usually takes care of the problem.

Making and Placing the Papercrete

Papercrete is an excellent wall covering for earthen materials, as it is nontoxic, recycled, good insulation, and local. It is sometimes called "padobe," and is a mix of repulped paper fiber – such as newspaper – and clay, sand or cement, and water. Although cement detracts from the sustainability of the material, it significantly strengthens the mixture. Portland cement– a material created by burning limestone that is unlikely to naturally occur on your building site – is one of the hurdles that green builders try to get around in many situations, but in this case, the amount of cement is fairly small and notably strengthens the mixture. Sand helps reduce shrinkage and is also fire-resistant.

Fire is a concern for those who are experimenting with papercrete, and it should be noted that mixes that contain less than a 4-to-1 ratio of cement (or earthen materials) to paper by weight might smolder or burn.

Papercrete is a moisture-absorbent material; keep this in mind when determining whether it is appropriate for the building site's climate. One factor to consider when using papercrete is whether or not you should seal it. Papercrete that is not sealed to allow moisture transfer can hold up longer than sealed papercrete in more arid environments. Should sealed papercrete absorb moisture that it is not able to release due to a sealant barrier, it could deteriorate considerably; in a dry area, sealing papercrete hinders its ability to release moisture.

On the other hand, papercrete can easily support mold and become weak if it remains moist for an extended time, so using silicone products for sealing papercrete can help – and can be a necessity – in climates with high amounts of moisture. Make sure the papercrete is completely dry before you apply any sealant, and the seal you create is continuous. At all costs, avoid trapping moisture under a sealant that has no way to escape quickly.

Depending on the size of the job, mixing the papercrete can be accomplished in large buckets using a hand drill with a paint mixer attachment, or a more mechanized process in larger tubs with mechanical mixers.

About 60 pounds of paper, 65 pounds of sand, between 60 and 90 pounds of Portland cement, and about 160 gallons of water (enough to achieve the desired texture) is a typical mix for papercrete. Some also recommend adding a shovel full of borax to the mix to keep bugs away. Here is how it is mixed and applied:

- Cut newspaper or other paper material into small bits and soak them in a portion of the water for at least one day.

- Mix up the paper and water thoroughly; the consistency will be very runny.

- Add sand and cement, and stir until the consistency is something resembling thick buttermilk.

- If using a machine mixer, which could significantly speed up the process, you may be able to skip soaking the paper beforehand.

- After the paper is thoroughly mixed, pour your mixture over heavy screen or wire mesh to rid of any excess water.

- Apply the papercrete, which should now be about the consistency of soft clay, directly by hand, and pat it into crevices, making the outside surface fairly even.

- The density of papercrete depends on the amount of earth or cement you use in the mix. A higher proportion of mineral makes a denser substance. Papercrete should be applied a few inches thick.

- Papercrete takes a considerable amount of time to cure, which lengthens according to the amount of paper used. Air circulation speeds the process, but humidity and cold temperatures decrease it.

Because sustainable home builders tend to choose papercrete due to its low cost, malleability, and sustainability, the number of informative resources available continues to increase. If you are interested in using papercrete in your construction, check out **www.livinginpaper.com**. Not only will this site offer you plenty of tips about papercrete and its many uses, it will give you comprehensive information about mixes to create structural papercrete blocks, several interesting facts about how papercrete performs, and the science behind the material.

Chapter 3

What is it Like to Build Your Own Home?

As you likely noticed in the previous sections, there are a number of factors that go into your average house that can require professional help — or, at least, professional consultation. However, the hardy owner-builder can often find his or her own ways to achieve goals by simply changing personal expectations of what a house really is.

There is no denying that a house is first and foremost a shelter, but modern homeowners — particularly modern American homeowners — are often set on making their house perform as much more than a shelter. If you need your home to function as a laundromat, a movie theater, and a day spa, there is a good chance you will be calling for a huge amount of professional help; if you simply want a roof over your head and a place to sleep, you can probably do it yourself for under $100 (think "tent" or "cave with a door"). Most alternative owner-builders fall somewhere in between these two extremes, and the side you lean toward will determine a lot when it comes to construction costs and the amount of professional help your project will require.

This following interview with Michael Blaha of **www.ilovecob.com** is probably the most striking depiction of what an owner-builder must go through to find the balance between what can be done on his or her own – and what cannot. It also illustrates very well the process – from inception to fruition – of constructing an alternative dwelling.

CASE STUDY: WHY HE LOVES COB

Michael Blaha, owner-builder
www.ilovecob.com

When did you begin your project?

May 31, 2007. [This particular house] is a low-cost alternative and natural building hybrid. It is off the grid, meaning its power comes from the sun; its water is harvested off the roof when it rains; and its heat—or some of the heat—is produced via passive solar design (heat from the sun with propane backup). The roof is pretty conventional, though the walls are mostly natural building materials. The foundation is earthbag on a rubble trench. The walls are all a variation of earth, sand, straw, and wood. Techniques used for the wall systems were straw bale, light clay, earthbag, and cob. Every surface has been covered with a generous amount of earthen plaster.

What first drew you to this style of building?

I have been attracted to organic architecture for a long time. Initially, my fascination began on a cob Web site, which had a picture of The Tsui House in Berkeley, California. It was erroneously placed on a cob page since it is largely cement; its whimsy was that of a cob home and its organic, ergonomic, and sculptural nature was to die for: Art you can live in? Sign me up.

CASE STUDY: WHY HE LOVES COB

Since then, I have been trying to get my hands on all kinds of materials for building and for sculpture. I experimented with everything that interested me, trying my hand at cob, straw bale, papercrete, padobe, earthbag, post and beam, wattle and daub, and ferrocement. I have gravitated toward materials that have the potential for high-energy efficiency, low cost, and low-embodied energy, and that usually comes with a high labor input.

Is your house complete? If so, when was it completed?

It is not quite complete, but it is habitable. Complete is a relative term. Does it provide the amenities of shelter? Yes. Are there a million things that still need to be done? Yes.

My recommendation: Do not move into your construction project. Those details that need finishing have a lesser chance of being finished when you live in the space. I hope to "move out" and have another go at finishing things, which just means turning my house back into a full-fledged construction zone: dust, mud, messy plasters, saw dust — you name it.

Where is the house located, and what kind of area was it built in?

It is in a rural area in a remote place, off from a remote town in northern New Mexico. It is a little too far out for my taste, but I am managing the adjustment.

How much construction experience did you have before you built your alternative house?

I have always enjoyed making things. When I was growing up, my family home was under construction; I must have gotten the gene from my dad. I had gained cob construction experience from a Cob Cottage Company (**www.cobcottage.com**) workshop, then went on to build Project MoonUnit, a structure made of a mixture of different eco-friendly materials. That was the bulk of my learning experience. I feel as though that is where I learned what not to do, and I gained the full cottage experience sans utilities. The learning continues.

Who helped you construct the house?

My friend Kevin helped on the house, and we took it on as a shared venture. There is nothing like having two able-bodied, able-financed, ambitious builders with a little too much time on their hands. We both brought different backgrounds to the project. The first part is a shared dwelling where we planned to share a kitchen, shower, and utilities. Since 2008, Kevin fell in love and moved to Idaho. So it is just me again: the solo-builder.

CASE STUDY: WHY HE LOVES COB

Did you employ any professional contractors during your project? If so, what parts of your home did they help you with?

We did seek professional services for the solar electric part of the equation because there were so many expensive components that required the respect of professional consulting — not to mention safety concerns. We needed to watch out when it came to electricity.

Did you have any professional design help for your home's layout? If not, what was the process of designing your own house like?

No, with the first part, it was mostly a, "What was easiest and quickest to construct?" design method: Make a box and create a space. It turned out OK. I might have changed my mind on the direct current (DC) water pump and the electric inverter fan if I knew how annoying the noises they made were at the design stages. It is fun bouncing ideas off someone during the design process; we had many creative "What if?" conversations all with in the shell of the space.

Are you happy with the finished product?

I have mixed feelings about it. It is not complete, so maybe I will feel differently when it is. There is quite a satisfaction inhabiting your very own dwelling. While unfinished, one must cultivate a presence of mind that can dismiss the things that are unfinished until it is time for action. I find that the little things that are "undone" bug me on a regular basis, but I am able to realize it all takes time. Most of all, patience is key. I am OK with how things are for now, but I do look forward to creating a quality, finished space.

What things do you think could have made the building process go smoother in retrospect?

Although flying by the seat of your pants is exhilarating for a while, planning — serious planning — would be helpful. I prefer the design and build method; there is nothing like being in an empty frame to truly get the sense of the space. You cannot get that on a sheet of paper or a computer schematic.

Savings is another large piece of the smoother process. I do not care if you are making a hut out of duct tape and chewing gum. You still have to buy some materials and spend money. I hate to say it, but if you do not have enough money, then do not build. At the very least, you must have the patience to do without some of the creature comforts or idealistic notions if you just do not have the budget to make them happen.

CASE STUDY: WHY HE LOVES COB

If you had to do it over again, would you choose to build a more conventional house?

I do not think so, but there may be some design changes I would have made — I would have created larger overhangs and maybe porch spaces. Perhaps I will try my hand at lime plaster for more durability in the future. I like the Earthship idea, where the bulk of the external skin of your house is earth; you do not have to replaster or repaint an earth.

If I were to do it again, I might want to put it on wheels; who knows. Of course, I probably would not be using cob and straw bale then. It would be cool to be able to move it around with you, for instance, if you decided you would like to spend the winter in San Diego or the summer in Minnesota.

What would you say was the biggest setback you experienced during the building process?

The biggest setback is that when you are building, you are not making money, and in addition to that, you are actually spending it. But you make your tradeoffs. One day soon it will all be paid off, and then there I will be debt-free. If you do not have bags of money lying around, then you become resourceful — more resourceful than you would have ever thought possible. At every turn and every step in the building process, you have the choice to spend, to salvage, to dumpster dive, and to do without for the time being.

What would you say was the most successful part of the building process?

I feel that the success is what happens within yourself when it comes to the building process. Have you ever been faced with such a large project that seems so huge that you have no idea how you are going to do it? Or, it could just be a number of small projects that add up into one big one. It is great to see it come to fruition, knowing that you did it.

However, it is a success that I am typing on a computer that is powered by solar electricity in a home, sheltered from the weather. I have learned so much throughout the process.

Was there any part of the construction that was easier or more difficult than you anticipated?

The devil is in the details, my friend. While creating a box with a door and some windows may be pretty straightforward, I cannot believe the time and attention the details required. The mechanical side of things seemed pretty tough at first but ended up being fun when you finally figure it out.

CASE STUDY: WHY HE LOVES COB

What were the reactions of your friends/family when you first began talking about building your own house? Did their opinions change during or after the construction?

Friends and family have been with me through the highs and the lows. I have a propensity to build things and make things with mud, so they understand my addiction.

What is your favorite room of the house?

Well, at 280 square feet, the "favorite room" is the room. I like that it is small and simple. My favorite feature is probably that it has passive solar heating, so it never seems to freeze and it gains free energy from the sun in the winter. It is nice and cool in the summer, too, with about a 25-degree difference naturally. I have always wanted to "make my own" energy, so I am happy that the power I use is created from the sun. It is such a different feeling.

How close does your house look and feel compared to what you imagined it would be before you began construction?

I always seem to sketch out fantastical spaces, though they never seem to turn out like that due to practical construction realities. We did not really have many preconceived ideas about what it would be like before hand. That said, it is a cozy, little space that does the trick. For me it is more about function than form — at least this time around.

As far as performance, how would you say the house you built compares to more conventional houses?

Even though it was not created with the utmost of efficiency in design, it still performs pretty well. Any house can be passive solar, so a conventional house could have the same benefits. As far as maintenance goes, I am still trying to get around the earthen plaster maintenance bit. Like I said before, lime plaster and design decisions could help with that.

Did your home cost more or less than you thought it would?

Much less — less than the conventional house, at least. It is almost what you might consider an "emergency shelter." But, you will not find many of the usual amenities of a conventional house. We started this project hoping not to invest too much. The land was cheap for a reason. I like trees, but there are not many growing out here. I like fertile soil, of which there is none. Just a sea of short forest (sage brush) here in the high desert mesa. Sometimes I wish I would have considered other sites and not jumped at the cheap land. The first thing you must do is to love your building site. Love it! Then tackle the building process.

Cordwood Homes

provided by Richard Flatau

The term "cord" is in fact a measurement often used when purchasing or preparing firewood; it refers to a stack of wood that is approximately 4-feet tall by 4-feet wide by 8-feet long. Cordwood building gets its name from utilizing fireplace-sized pieces of round or split wood that might otherwise be seen as refuse. If these pieces are in good condition and of an appropriate type of wood, the savvy alternative builder can do a lot more with this material than burn it or throw it away.

Wood is not the only component in a cordwood wall; it has been estimated that the average cordwood wall actually comprises about 60 percent wood and 40 percent mortar. Typically, masonry mortar similar to the type used in brick and stone work is employed to cement the individual pieces of wood together, creating a unique aesthetic and a formidable, energy-efficient barrier. While mortar or concrete is an excellent and practical choice for filling in the gaps between the logs in a cordwood structure, the inclusion of Portland cement has prompted some cordwood builders to opt for cob as an alternate mortar material. Using cob instead of traditional cement mortar will probably be less expensive — and can be considered a more sustainable practice — but it will also be more labor intensive and may not be as moisture resistant.

Whether mortared with cob or concrete, cordwood homes comprise some of the most beautiful houses in the alternative building world. Their look is unmistakable, and when properly constructed, their function is second-to-none. A cordwood shell is an excellent insulator, and the structure itself will easily last a lifetime with very little required maintenance. Though a lot of preparation time is necessary to do this type of construction right, once your logs are stripped, cut, and aged and your mortar is ready to be mixed, a cordwood wall can be constructed rather quickly.

About Cordwood Construction

provided by Richard Flatau

When considering cordwood construction, it is wise to first take a good look around the property on which you plan to build. If you see nothing but grass and shrubs or sand and cacti, you might want to think about a different style of building. The main ingredient in cordwood construction is, of course, wood, and if you do not see any wood that is easily accessible on or near your property, there is likely another material that may be easier and cheaper to access.

Harvesting cordwood suitable for building can be done in several ways. If you are building on land that is surrounded by forest, you can fell your own trees and create your own cordwood. In some cases, you might be able to get all the trees necessary

Photos from Alternative Home Builders

all cordwood photos provided by Richard Flatau

Cordwood Lodge

Bottle-End Wall

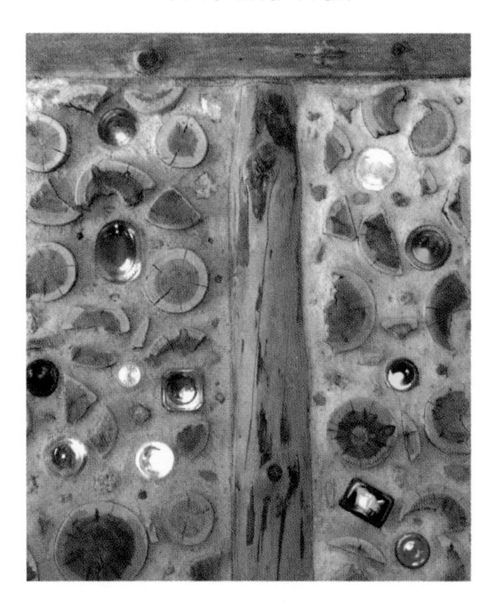

Stackwall Corner

Photos from Alternative Home Builders

all adobe, sod, and cob photos provided by Michael Blaha

Adobe Roof

Detail of Adobe Wall

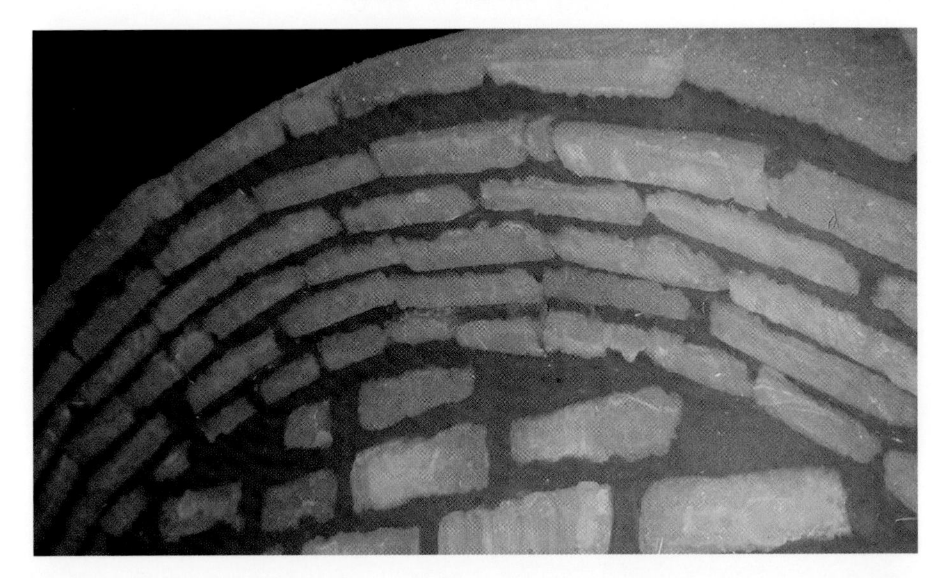

Sod Home

Sod Interior Wall

Sod South Front

Cob Home

Photos from Alternative Home Builders

all earthship photos provided by Earthship Biotexture

Corner Cottage Entry

Corner Cottage Porch

Water Catch Off Roof

Detail of Recycled Metal Roof

Hybrid Earthship Solar Panels

Phoenix West in Snow

Detail of Corner Cottage Porch

Detail of Mud and Bottle Wall

West Water Catch

East Entry of Earthship

Corner Cottage Master Bedroom

Tub with Bottle Wall

Modular Earthship Arches

Bottle Detail in West Bedroom

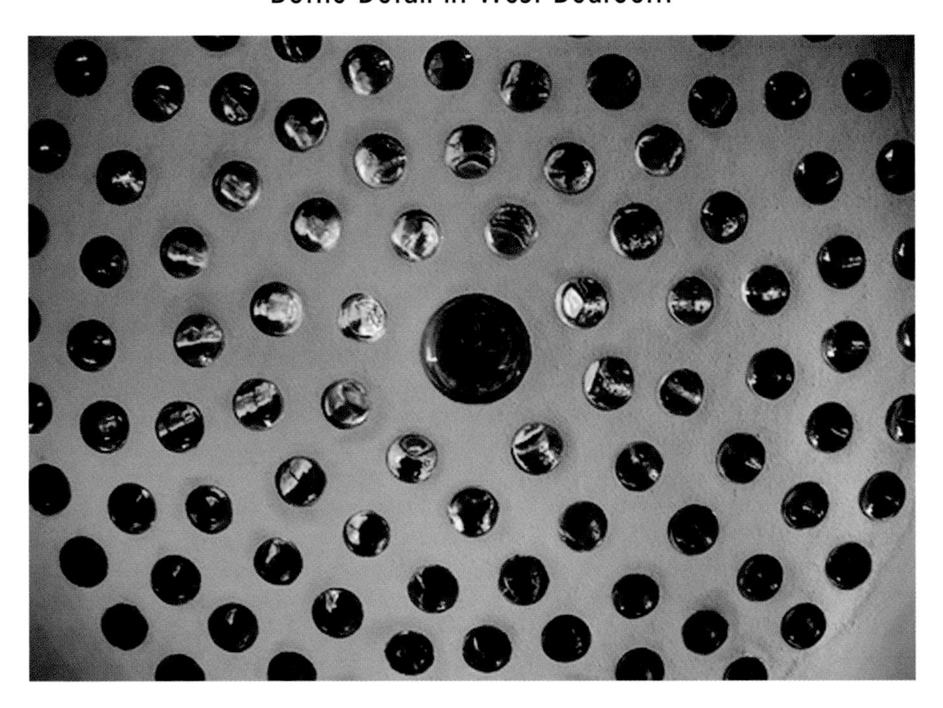

East Bathroom Bottle Detail

Bottle Entry

Solar Gain

to complete your construction simply by clearing the site you will be building on. Cordwood can also be purchased relatively cheaply under the name "firewood." If purchasing firewood, find out what kind of tree the wood was cut from and how long that wood has been aged. Different species of wood draw and release moisture at different rates and experience different amounts of swelling and shrinking when exposed to heat or moisture. Due to this fact, it is generally recommended that you use only one species of wood in your construction, which can sometimes be difficult to do when purchasing firewood.

You may also be able to harvest wood from trees that have fallen naturally. This is certainly a solid, green, and sustainable practice, but finding trees in good shape that have fallen naturally is a matter of luck rather than planning. Standing dead trees and trees that have fallen but are propped up by nearby branches — not touching the ground — are often excellent sources of naturally occurring, non-rotted wood. If already-fallen trees cannot be found nearby, consider using wood that might be cut down on the site of someone else's future home. You may also be able to procure fallen trees collected in urban forestry programs for cordwood construction. As a sustainable building product, felled trees treated as waste and trees cut down from the home's building site are a better option than cutting down trees in another area. If cutting trees from the site, then the less impact to the environment, the better. Ideally, a cordwood home can be built using only those trees that need to be cut down to place the home, but if this is not possible, smaller trees can often be harvested from other places without having as much of an ecological impact as cutting down larger, older trees.

It is necessary to age wood before it is used in construction. "Green" is a term that is frequently used to describe a freshly cut tree. Green wood is full of moisture, which can be a problem when that wood is used for construction. Logs naturally expand and contract depending on changes in temperature and moisture, and green logs are about as full of moisture as they will ever get. Aging a log reduces the amount of moisture it contains, which actually shrinks the diameter of the log. If this shrinkage occurs after a log is already a part of a wall, gaps will be created around the log. In extreme circumstances, that log can actually become loose. It is better to allow this process to take place in a barn or shed rather than in the wall of your house. Obviously, aging done in a dry place tends to be quicker and more effective than wood aged in a field. Wood that is simply left outside should be covered on the top to reduce exposure to precipitation, but if you have access to a shed or barn, this is a good time to use it.

provided by Richard Flatau

provided by Richard Flatau

While aging is a good idea, you do not want your logs to undergo the process for too long. Though rare, there have been cases where extremely dry logs used in cordwood building have actually expa-nded enough after being set in a wall to crack the mortar used to hold them together. As you can likely imagine, this compromises the stability of the overall structure; in-wall expansion is, therefore, a far more serious problem than in-wall shrinkage, which can often be solved by simply filling the gaps around shrunken logs. Expansion is, again, pretty rare, but is a serious enough problem that it is worth taking the necessary steps to avoid the risk. Expansion problems are most often associated with hardwoods that have been aged for a year or more; by using softwood that has aged for about a year, or hardwood that has been aged for three or four months, you can eliminate any chance of in-wall expansion.

Cordwood suitable for construction should be free of bark. If you are felling your own trees, it is best to remove the bark as soon as possible. Due to the annual fluctuations in tree biology, spring is generally the time of year when bark is most easily removed. It is also preferable to remove bark soon after the tree is felled; if at all possible, bark should be removed the same day the tree is cut down. The longer you wait, the more difficult it is to remove the bark. Failure to remove bark will increase the wood's necessary aging time and make the wood more prone to insect infestation and rot – during the aging process and after construction is complete, too. Removing bark is not an extremely difficult process, but it can be tricky at first. There have been reports of people debarking a freshly cut tree in the spring time with nothing more than a sharp knife and their fingers; however, to be safe, use two tools to help make the process easier: a heavy, sharp tool (like an ax or hatchet) to slice the bark and a sturdy, flat tool (like a steel trowel) to slide underneath the bark and pry it away.

When choosing wood, be as certain not to use insect-infested or rotting wood, as it can cause serious structural problems if included in construction. No matter where you obtain the wood from, make sure to check under the bark before hauling it to wherever it will be aged. Rot and insects beget more insects and rot, and bringing home a serendipitously found felled log that is secretly swarming with bugs, fungus, or rot can be disastrous if it spreads to the wood you already have stored.

Cordwood construction can use either round or split logs. Using round, unsplit logs will get your walls erected faster because you will need fewer of them to achieve the desired height. Split logs age faster and generally have fewer problems with shrinkage but are more time consuming to build with and labor intensive to create. Both techniques are acceptable in most situations; the final decision to choose split or round logs is often more of an aesthetic choice than a practical one. There is also no reason why you cannot do a little of both.

The logs you use to construct a cordwood wall should be about the same length; commonly, logs used in cordwood construction are between 14- and 24-inches long. Longer logs are better insulators, and wider walls tend to provide more stability. Thus, the length of the logs you use depends on the climate of where you are building and whether you need your walls to be load-bearing. It is also important to be sure that a log's dimension is fairly consistent from one end to the next to help with stability within in the wall. The diameter of your logs can and should vary from one log to the next, as this not only adds to the aesthetic appeal of the finished structure, but will also allow you to create a snugger log structure.

provided by Richard Flatau

Unlike many other types of alternative construction, cordwood structures are a perfect fit in cold climates. Of course, the thicker the shell of your house, the more protection it will give you from the cold. Any properly constructed cordwood wall will have decent thermal mass and provide significant insulation. Softer woods are lighter, less dense, and naturally contain more air pockets than hardwoods, which makes them better insulators. Softwoods can have an insulation value of around R-1.25 per inch. Hardwoods provide better thermal mass because of their density, but their insulation is only about R-.09 per inch.

A variety of woods are considered better than others because of their resistance to rot: juniper, cedars, Pacific yews, hemlock, and bald cypress have the least likelihood of rotting, while spruces, pines, poplars, and firs also work fairly well. Notice that most of these rot-resisting woods are softwoods: Because of their lower density, they are also less likely to expand and contract due to moist conditions. Regardless of the type of wood, always include large overhangs on the roof to protect from water damage. You may also consider plastering over the wood to increase its durability and reduce air infiltration due to wood shrinkage – though many consider this an aesthetic travesty because exposed cordwood walls are often considered attractive.

Some cordwood builders choose to seal the log ends to protect the wood from moisture, rot, and insects over the life of the home, but many view this as a mistake. Cordwood, like other natural building materials, is meant to breathe and adjust to changes in temperature and humidity, and sealing the log ends hinders this. Additionally, moisture that makes its way into your logs should have a way to escape; by sealing log ends, you are blocking off the escape route of internal moisture, which could lead to problems with water retention and decay.

CASE STUDY: DOING IT ON YOUR OWN

Tom and Susan Nunan
Owner-builders

Tom and Susan Nunan's journey in alternative construction began when a relative found an article in a magazine about cordwood building in winter 1994. This got Susan's wheels turning. After reading the article and a few books on the subject — which all claimed that you needed little to no building experience to construct such a house — the Nunans decided it could work for them. While the actual construction of the house did not begin until the late spring of 1996, preparations were underway to build a cordwood house on a plot of land that had been in their family for decades. "On a bitter cold, dark, stormy day," Susan said, "we contacted a lumber company. A bearded man met us at the door and led us into a small room with a wood-burning stove for heat.

There were two other bearded men there. They knew everything there was to know about wood — they had a big discussion over ash versus red pine; red pine won — and took our project seriously."

While many cordwood builders must strip their own logs of bark, these men felled the trees and stripped the logs for the Nunans for a reasonable price.

While Susan described their cumulative construction experience before building this house as "none," Tom preferred the term "moderate." Wherever the truth lies, it is clear that this undertaking was both formidable and intimidating.

"I had no idea how big a house is until I started building one myself", Tom said. "In the beginning, when all I had was a flatbed load of materials and a concrete slab, it felt like I was trying to build the Empire State Building. I called it 'the weekend project from hell.'"

They kept it simple. While there is now a room behind a bookshelf and a small deck added on to the second story, Tom made the original design without frills. "We were lucky with the permits," he said. "The code officer was an older man who was also the local barber. I think he took a fancy to my renegade, survivalist-style architectural drawings on two 8.5- by 11-inch pieces of graph paper."

Professionals put in the septic tank and well, and a relative's construction company poured the slab foundation. Once these steps were completed, the Nunans held a party where friends and family showed up to lend a hand; many of these people came back time and time again to pitch in for a few hours or a few days.

CASE STUDY: DOING IT ON YOUR OWN

Susan claimed that people kept showing up because the work made them feel good. She has even gone so far as to describe the building of their home as "therapy."

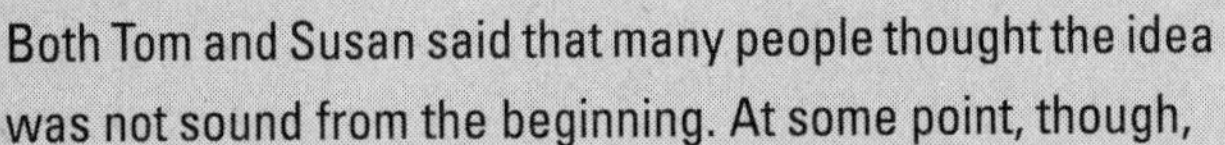

Both Tom and Susan said that many people thought the idea was not sound from the beginning. At some point, though, the house began to take shape, and enthusiasm grew. Those who were skeptical at the start eventually came around, and those who were staunch supporters from the beginning were elated whenever they stopped by the job site.

After six months — three more months than anticipated — the house was habitable. Tom affirmed that the experience of building his own house was a good one, despite some minor bumps in the road.

"Things went amazingly smoothly. It was all a part of the divine plan for this corner of the Milky Way," he said, though he admits that a few things could have made the process even smoother.

"Bigger muscles," he said, "and hands like mitts."

As for completion, neither Nunan is comfortable with the term.

"It's mostly complete," Susan said. "I'd like some kitchen cabinets and to have the ladders removed from around the outside."

The couple is always doing something to improve upon the house. "Maintenance is simple when you build your own house," Tom said. "I know how everything is put together".

As for performance, Susan said, "Our electric bill is about half of what a comparable square-foot house is, and since we heat with coal and wood, our heating bills are a fraction of conventional houses. [But,] when it's sub-zero and the wind blows, you have to wear your coat inside."

In the end, both are proud of their home and love to show it off, and they also consider the total price of their home a bargain.

Constructing Cordwood Walls

provided by Richard Flatau

As with any other type of construction, there are a couple different options when it comes to building a cordwood wall. The option you choose will have a direct effect on the wall's function and load-bearing capacity, so deciding early is a must.

The mortar that goes between the wood in a cordwood wall can be continuous from interior to exterior. This method is generally not the preferred one for two main reasons: It costs more, and it creates a less insulated wall. In most cordwood construction is a gap between the mortar on the outside of the logs and the mortar on the inside of the logs. The empty cavity where no mortar is present is generally filled with sawdust or another cheap insulator; this means that you will use less mortar in the overall construction, which can save you a bunch if you are using a mortar that includes cement in the mix and have less heat transfer between the outside and your living space, as well.

Cordwood walls can stand on their own, or they can be created in between the posts within a post-and-beam frame. In the former, the walls must either be curved or built-up at the corners in order to be considered load bearing. In the latter, the posts rather than the cordwood and masonry are responsible for bearing the roof load. Neither option is better for every situation. Building curved or built-up cornered cordwood walls tends to make the overall process faster as you skip the framing step; building within a post and beam frame gives you the benefit of being able to erect a roof early, which gives you some leeway with Mother Nature during the building process. If you are building on a concrete slab foundation, creating a load bearing cordwood shell might be easier; if beginning from a pier foundation, non-load bearing cordwood walls built within a post-and-beam framework might be easier. Some like the look of one or the other better, which seems a good enough reason to go with it. Just make the decision early so you know how to proceed.

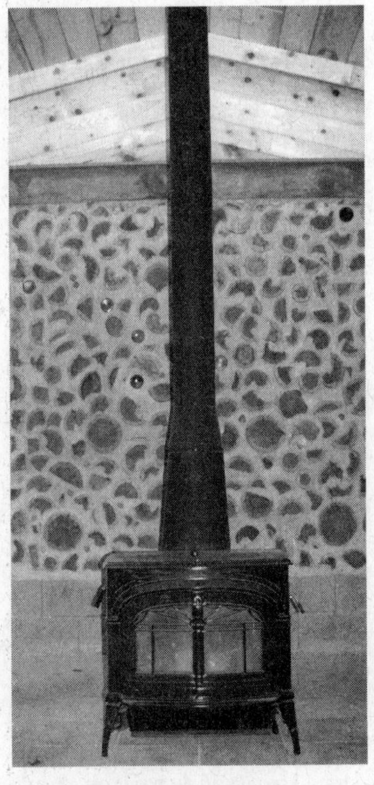

provided by Richard Flatau

For load-bearing cordwood walls, the longer the logs, the thicker the wall will be, and — theoretically — the stronger the wall will be. When dealing with cordwood walls constructed within a post-and-beam framework, the length of the logs you use is typically determined by the thickness of your posts. The only reason for your logs to be any thicker than the posts they will be set between is to provide a higher level of insulation; using longer logs in this scenario may look shoddy, however, and because they are not providing structural support, they are not making your wall any stronger.

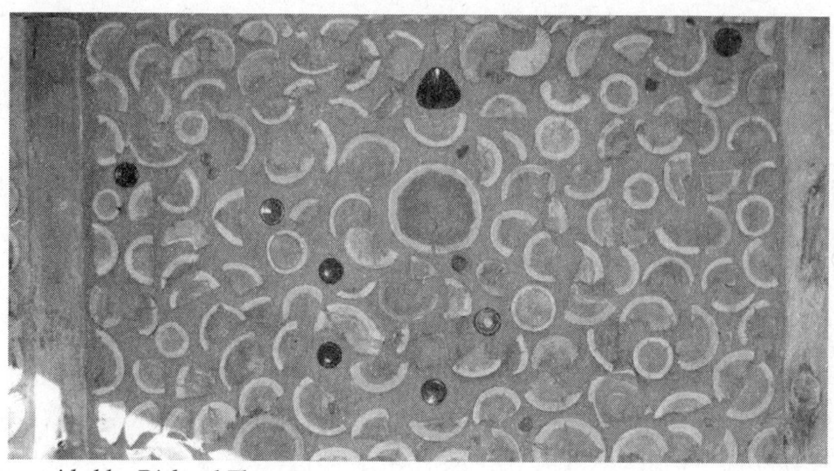

provided by Richard Flatau

All cordwood walls depend on two primary components: the wood and the mortar. Mixing mortar is a job that is hard to pick up at first, but one you will likely improve at it. All mortar consists of dry components mixed with water. Those who have done this job before will have his or her own individual recommendations as to the most effective dry mix: Some will say claim one combination is perfect, while others will claim that they have tried that mix, but prefer another hands-down. By the end of the process, you will form your own preference.

An "S-type" masonry mortar, notable for its combination of compressive and bond strength, is made using two-parts Portland cement, one-part hydrated lime, and nine-parts sand. This is not always the best mix for every situation, but it a good recipe to start and tinker with. Many folks who deal with cordwood masonry on a regular basis also recommend adding a portion of moist sawdust to the mix as both a binding agent and a curing retardant.

If you are building non-load bearing cordwood walls within a post-and-beam frame (see the framing section of this book), the entire frame and possibly the roof will be in place before you begin doing any cordwood work. Once the frame is complete, simply fill in the gaps between the posts with cordwood and mortar.

If building a load-bearing cordwood shell, the built-up corners are the place to start. These corners are the key to making your cordwood shell suitable for bearing the load of a roof or a roof and second floor, so it is important to do them correctly. You can use round logs to create built up corners, but this is generally not considered ideal. Round logs have been used in built-up corners and can perform effectively; however, if you have access to logs that are squared on two or three sides, these will provide the best stability. Once the corners are erected, plumb, and level, you can stretch several lines in between them to serve as a guide for the walls they will be connected by. These lines can be made of any visible string or twine, and will take a lot of the guesswork out of keeping the connecting walls straight and true.

Whether your logs are round or split, your walls are load bearing or non-load bearing, or your wood is hard or soft, the process of laying logs and setting them in mortar will be a lengthy but ultimately fulfilling process. Even those planning on using round logs for the majority of the house often begin the lowest course of cordwood with logs that are split in half, with the flat side facing down. Each course of logs should, at the very least, be staggered and not line up with the row below it, but if you are using logs of dramatically different diameters, the process will be less about making rows as it will be about finding places where a particular log will fit.

provided by Richard Flatau

provided by Richard Flatau

Chapter 4

Adobe and Light Clay Homes

As the avid alternative builder knows, you can do a lot with earth, straw, and water. This book will discuss three commonly used alternative building materials that are made up primarily of these ingredients. Adobe — the best known and most popular of these materials — is the eco-friendly man's brick. Its benefits are numerous and its uses widespread. On the other end of the earth, straw, and water spectrum lies light clay. Often used as insulation rather than building material, light clay is not as structurally strong as its adobe cousin, but its versatility and insulation value make it a material worth a second glance. Cob, the third material made of these ingredients, will be discussed in its own chapter.

Adobe and light clay are generally considered warm, dry climate materials, but if properly planned and sealed, it is certainly possible to use these materials in a less optimal environment. In very cold areas, a simple adobe wall will not be enough insulation to keep a home's interior comfortable, so another form of insulation will be required. In very wet environments, special care in the planning stages – or frequent maintenance once the home is constructed – will be a necessity to keep exposed adobe walls in good shape.

About Adobe

Adobe is one of the most common earth-based building materials on the planet, concentrated throughout many of the drier areas of the globe; additionally, it is one of the oldest types of building material still in existence. Because adobe's use is so widespread and its history so long, it is a material that many people first think of for their alternative housing. Whether used for the construction of walls, floors, or roofs, adobe can be an economically and environmentally accessible material.

Adobe has existed for thousands of years. The Egyptian word for adobe, which means "sun-dried brick," extends back 4,000 years to 2000 B.C.; this word morphed into "tobe" for "mud brick" by about 600 B.C. Cultures ranging from ancient Egypt to Medieval Europe and from China to West Africa used the material to construct homes, churches, mosques, and walls. Adobe was used even earlier in Mesopotamia. Each region's adobe structures are unique, of course, because of the regional topographical and soil differences. Adobe structures have been commonplace in the American southwest for hundreds of years and have been built by Native Americans, Spanish settlers, and modern builders today. The hot, dry climate makes it particularly well-suited to this type of building, as adobe homes are generally naturally cool inside. Areas where heavy rains or cold temperatures are the norm are not considered prime locations for adobe construction.

Made from what is essentially mud, adobe homes are sturdy structures constructed of bricks that are up to 30 percent clay and about 70 percent sand. The clay and sand are mixed with water and often strengthened with a material such as straw. This mixture is then shaped by hand or poured in a form to create adobe bricks that are baked in the sun's natural heat until stiff and sturdy. Some builders even skip the process of shaping the adobe into bricks, preferring to instead shape it directly into the wall structure. However, this second method is less common than the tried-and-true process of forming individual adobe bricks to build a structure. This process is fairly labor-intensive, and it takes time to form the many bricks and let them cure; some adobe builders spend a year just creating the bricks. However, once completed, adobe homes are attractive, long-lasting, and very green structures that, in spite of the lengthy prep time, are relatively inexpensive to build.

In most cases, adobe is not used as a load-bearing material. Both stick framing and post-and-beam framing can be used to provide the load-bearing components of an adobe house. In the rare cases where adobe is used to bear the weight of a second floor or roof, special attention must be taken to ensure the overall stability of the structure; the help of a structural engineer may be necessary to successfully construct load-bearing adobe walls.

When the modern homeowner thinks of a brick, there is a good chance he or she is imagining a relatively compact, red piece of material. Adobe bricks can certainly be made to conform to this description, but because they are handmade, it is generally more efficient to create a fewer number of larger bricks than a larger number of bricks of conventional dimensions. Imagine a brick that is about 4-inches thick, 14-inches long, 10-inches wide, and 30-35 pounds, and you will have a better idea of a typical adobe brick. These dimensions are not the rule; you can make your bricks smaller or larger if you want. However, the described dimensions create a brick that most owner-builders appreciate: large enough to create efficiently, but not large enough to be particularly strenuous or cumbersome to work with. Make no mistake: No matter the size of your bricks, this job will have you exhausted at the end of the day. Fortunately, when the job is over, you will have a home – and likely a few pounds of new muscle – to show for all of your labor.

The structure for adobe must support the bricks, which are stacked into thick walls that sit on top a foundation, such as concrete or stone. Typically, the same mix used to construct the bricks is also used as mortar in adobe walls, but this time, it is wet.

Bricks, the main ingredients in the house's structure, serve as a substitute for wood or cement in the walls, which can be costly and are far less eco-friendly. Also, because the material used to make adobe is typically local, fuel and other transportation costs are greatly reduced. Some builders add components to the adobe for reinforcement in excessive weather, such as concrete or emulsified asphalt. Though adobe homes are essentially sustainable, the addition of emulsified asphalt detracts from the home's otherwise environmentally sustainable character; asphalt is a petroleum-based product that produces gases, which can be toxic. The climate and building materials used in construction have sometimes made emulsified asphalt a practical addition to your adobe mix, but if it is not a necessity, the green builder will be better off without it.

The energy efficiency of adobe's thermal mass can be quite high. The thick-walled earthen material helps maintain fairly even temperatures, particularly in a hot, dry climate, keeping it cool in the summer and warm in the winter. Although the adobe bricks do not have to be insulated, the walls sometimes require a double wall constructed with an air-filled cavity, or some other type of insulation. In climates with temperatures that dip near freezing for an entire day or longer, adobe walls will certainly require some form of insulation to perform well. Adobe technically does not have an R-value, but some consider it to be less than R-1 which is simply not enough insulation for cold climates.

Adobe's "thermal mass," or heat-storing capacity, helps make rapid outside temperature fluctuations less noticeable inside, making it ideal in desert regions where temperature fluctuations over a day can be quite extreme. Adobe, cob, and other thick earthen materials are better at passive solar heating and cooling than many conventional, and even other alternative, materials. In addition to decent energy efficiency, adobe walls regulate humidity in the home as the material soaks moisture from the air and releases it back into the house. Adobe is also good at reducing noise and is naturally fireproof. Overall, in a warm, dry environment, adobe is an excellent building material and is often easily accessible.

As far as providing a healthier environment in which to live, adobe's nontoxic materials do not carry the allergens and toxins that conventional home building materials carry. People who suffer from asthma and allergens, and are sensitive to chemicals and VOCs, can find solid relief in an earth-based material like adobe.

Once a person is comfortable working with adobe, it becomes easier to understand how adobe's resilience and functionality make it well-suited for several facets of the home construction process in addition to the actual bricks and walls. Adobe flooring can be used for a home's foundation; adobe plaster can be used to bind the adobe bricks together for construction; and adobe mortar can be used to fill the home's many joints and bind bricks together.

For long-term durability, adobe design should allow for some method of protection for the walls. Building sites on relatively high ground, which are not subject to flooding or standing water, are optimal. Roofing with long eaves to keep moisture from hitting the bricks during a rain storm, and stem walls that keep the adobe from coming into contact with water flowing on the ground, can also be beneficial. If possible, avoid running plumbing inside adobe walls to avoid condensation or leaks that could compromise the bricks. Try and route pipes through conventionally constructed interior walls or tastefully on the surface of adobe walls.

About Light Clay

Light clay, alternatively known as light straw or straw-clay, is a variation on adobe and cob. While adobe and cob are more earth than straw, light clay is primarily straw that is coated with a mixture of near liquid clay, called clay slip, so that it will easily stick together. Light clay is most often used for interior walls to boost the insulation.

Though it is not used as a load-bearing material, the presence of light clay can be appropriate on the outside of walls to boost insulation or protect a material underneath. Additives such as linseed oil can increase light clay's resistance to water and rotting. Working with light clay is comparable to working with cob, except that light clay is lighter. Some builders adapt light clay to cob houses in areas in which the cob's structural support is not required. The beauty of working with straw and clay is the mixture's ability to be altered to fit the need of the situation. When necessary, you can simply change the ratio of straw to clay to fit the situation, but be sure to know the home's structural needs before adapting mixtures.

As with locating straw for other earth-based building needs, be sure to find clean straw. Farm communities often have some type of network for finding supplies, and asking around can often turn up quality straw.

To make light clay, follow these steps:

1. Soak a quantity of clay in water and thoroughly mix the two until it forms a runny, non-lumpy mixture (commonly referred to as a slurry or clay slip).

2. Lay out straw loosely on a platform or on a spot on the ground.

3. Pour the clay slurry onto the straw and mix. Turn the straw to coat it well with the clay mix. A good light clay mixture should resemble dark, dirty straw.

Light clay is best applied when wet. It can be stuffed into cavities — like stuffing a turkey — but to get the most bang for your insulation bucket, it can be applied in levels between wall studs (where rolled or batt insulation would normally go) before you put up drywall or paneling. The process goes like this:

1. Pick a pair of wall studs where you want insulation between and start at the bottom by nailing a wood board between the two studs; this board will act as a temporary form.

2. Stuff the light clay into the area created behind the board, packing it thoroughly.

3. After the clay mix has just begun to dry, remove the board and move it up to the level where the first area of light clay ends, and nail the board again between the two studs.

4. Stuff and pack the next course of light clay into the area created behind the board.

5. Continue this process to the top of the wall.

6. Once you hit the top of the wall, stuff the light clay as best you can into the space between the last course and the top of the wall; you might try nailing your form over this last course to keep it in place as it dries.

7. Light clay shrinks as it dries, so builders stuff loose straw into the gaps that form after the mixture dries on the wall and before adding plaster.

The builder need not wait for each layer to dry before adding the next. When done properly, the material is strong and densely packed, ensuring it does not slump once the temporary form is removed. Though the coated straw may not seem strong, the mix becomes a surprisingly tough material as the clay dries and binds the straw. Once it is completely dry and any shrinkage gaps have been filled with loose straw, finish the wall by installing drywall sheets or by applying plaster.

Adobe Brick Considerations

Regional location is an important factor to consider when planning the location for an adobe home. Though homes made from this material can be built in many locations and climates, their heating and cooling properties are most effective the closer the home is to the equator. Farther from the equator, the heat that is lost through the adobe walls is a powerful deterrent. Additionally, adobe tends to work better in drier regions, as areas with high amounts of rainfall create additional challenges for withstanding the burden of heavy moisture absorption. Other methods of earthen home construction, such as rammed earth and cob, work better in wetter climates.

Most adobe homes built in the United States today are single block, not double, and reside in the Southwest, which is particularly well-suited for adobe homes because of the cohesion properties in the clay-like soil. For those fortunate enough to be building on a site that has the right consistency of soil to create adobe, one of the first steps is to decide whether the soil will work by performing a simple soil test.

The material for producing adobe bricks will be a soil with the right mixture of clay and sand. If after testing the soil you find that it is lacking, you can add clay or sand in your adobe mix, depending on which element the earth does not provide on its own. Straw is then added to this balanced mixture — not necessarily to strengthen the bricks, but to bind them together and allow them to shrink and expand without the worry of cracking.

Knowing the soil's pH level will additionally help you decide whether the soil is sufficient for the job. This can be done by sending a small sample to the local extension agency or to a soil-testing firm. If the results show that the pH is low, with acid of 5.5 or less, then there is much organic matter in the soil, and it is not sufficient enough for adobe. Be sure to check the pH level in various areas of the property before you deem the soil unusable or you start trucking in additives supplement it. Soil composition can vary widely even within a small space, and testing different areas is the best way to find good, usable dirt.

For the test, find a clear jar with straight sides and a lid that seals tightly. Fill the jar half-full of the soil, then add water until it the jar is about three-quarters filled.

- Add a pinch of salt to the soil and water mixture, cap it tightly, and shake it until it mixes thoroughly into a single solution.

- Let the solution soak for a few minutes and shake the mixture again to ensure it is well-combined. Layers of particles of different sizes will separate as it settles. This part of the process may take a few minutes with salt added, or can take a few hours without.

- When the layers have formed, large sand particles will have settled at the bottom of the jar. Above that layer is silt that has settled out of the mixture, and a layer of clay on top. To determine the approximate percentage of the soil in each component, use a marker, pen, or tape to mark the height of each layer.

Once the right balance of sand and clay has been determined — typically about 50 percent sand, 35 percent clay, and 15 percent straw — consider whether a stabilizing agent will be used. If emulsified asphalt is used as a stabilization agent, your adobe mix is considered fully stabilized when the emulsified asphalt makes up about 5 percent of the total mixture. If using emulsified asphalt, it is important not to use a sealant on exterior walls so that you allow water vapor to escape.

Adding an emulsifier to adobe homes in the southwestern United States is generally unnecessary when plastering over bricks because the bricks themselves will not be exposed directly to the weather. However, an adobe home without an outer plaster will likely resist moisture better with the emulsifier.

Making Adobe Bricks

Ensure that the bricks are solid and will last through inclement weather by checking the guidelines in the International Building Code (IBC), which is a set of standards that many building certifiers follow. After determining the correct consistency and strength of the adobe mixture, the next step is to build a mold for the brick. It is necessary to have a mold in place, ready to form the brick, before mixing the mud. The mold can be made of wood, metal, or any other strong and sturdy material. You can, of course, construct a mold that is meant to create a single brick, but it is generally more efficient to make a mold that will form multiple bricks at once.

Whatever dimension you want your bricks to be, the inside of the mold must be equal to these dimensions. Keep in mind, however, that the mold will be taller in height than the final adobe brick product, as it usually falls or slumps a little when taken out of the mold. It is entirely up to the builder whether the bricks are or are not of uniform size. Slight variations in the bricks' sizes, shapes, and heights can produce visually interesting structures that are as sound as those made with perfectly uniform bricks if the material used to connect the bricks — the mortar — balances their discrepancies. Most importantly, figure out the home's dimensions and divide them by the size of the bricks for a general idea of how many blocks the house will require.

If making a mold from wood, always choose a hardwood, fir, or pine, as these will withstand the wear and tear of brick making. One step the builder should not neglect is treating the inside of the mold so the brick is easy to remove. Use paint, shellac, or some other smooth agent on the inside of the mold to produce a slick surface. If the surface remains rough, it will make removal difficult, adding time and frustration to the task. One trick is to line the inside of each square with malleable sheet metal that is easy to maneuver and shape and will let the finished product slip out easily.

Some adobe builders create a simple structure that resembles a ladder or a rack that allows for making multiple rectangular bricks simultaneously. Four or five squares are typical for these structures, without making them too cumbersome or heavy for builders. Another option for a mold is to make a simple, square box in which to pour the adobe mud.

A typical mold makes four bricks, and bricks can range in size but should not be too large to handle. The following measurements are most common for adobe bricks:

- 9 by 18 by 4 inches
- 8 by 16 by 4 inches
- 10 by 14 by 4 inches

Once you have your molds made, with interior dimensions that look something like those listed above, fill the mold with the mud, level the mud with a 2 x 4 or other straight object, and let the mud slightly harden inside the form. After a few hours of drying, remove the mold and let the bricks dry completely.

Making bricks by hand can be a long and arduous process. There are machines that can be used for creating adobe bricks, but their main benefit is a stronger brick rather than a faster process. If a machine press for casting bricks is preferred, the CINVA-Ram — a well-known device developed by the Inter American Housing and Planning Center in Bogotá, Colombia — is one that many alternative builders have used successfully. Slightly resembling a large garlic press, the CINVA-Ram is a light, portable adobe press that is simple to use and creates more uniform bricks for adobe construction than hand-poured bricks. Because its bricks are made under pressure, they are almost twice as strong as cast bricks. However, making bricks with a press is slower than casting with a form.

Methods of mixing include the very primitive and basic, to the high-tech and more expensive. Some professional alternative home builders recommend purchasing a contractor's plaster mixer to create adobe mud. The plaster mixer is a large gas-powered machine with blades, which contractors use for mixing mortar and plaster in conventional house construction. Alternatively, some people consider using a cement mixer, but it is usually not as efficient at thoroughly mixing the adobe mud. If using a professional mixer, look for plaster mixers at equipment-rental agencies. If purchasing a plaster mixer, used ones can sometimes be bought from an equipment renter or other professional firm, such as a masonry company or contractor. You may also have luck finding a used plaster mixer online.

To buy a plaster mixer or just browse options, the below Web sites offer different machine, in a wide range of prices, complexity, and user-friendliness:

- MBW (**www.mbw.com**) specializes in compaction and concrete construction equipment.
- Discount Equipment (**www.discount-equipment.com**) has a wide array of both new and used equipment for sale or rent.
- Southern Tool (**www.southern-tool.com**) sells a variety of construction tools and equipment and has a large selection of mixers.
- Canoga (**www.canogamixers.com**) offers a catalog of mixers in varying degrees of complexity and price.

After locating the best equipment for your budget, skill level, and staff size, and before launching into mixing materials, compile your equipment and materials into one area near the building site for easy access and speed. Gather a pile of dirt, the bale of straw for strengthening, the emulsified asphalt (if needed), and locate a place to combine all these ingredients, whether with machinery or by hand. Be sure that the brick-casting area is near the area for mixing the adobe to avoid extra work and to get the mixture into the correct form as quickly and easily as possible.

For drying the bricks, use an area that is flat and even. Cover the ground with dry sand or straw so that the adobe will not mix with the dirt beneath and lose its shape.

Mixing with a machine:

1. First, add 7 to 8 inches of water in the bottom of the mixer.

2. Shovel the clay on top.

3. Mix the ingredients, adding water and clay as needed to create a stiff mud. Work out all the lumps and clusters. This mixing will soak the clay in preparation for the next step.

4. Add equal parts of sand to the clay already in the mix. At this point, some builders prefer to mix the soaked clay and sand by hand — or by foot — instead of by machine as the mixture becomes dense and heavy. Using a mechanized mixer for this step depends on the strength of the mixer.

5. If using an emulsifier, slowly add the agent according to the directions.

6. Let the mixer work for a couple of minutes until the mix is thoroughly blended, then add the straw. A straw-to-mud ratio of about 1-to-5 is an approximate goal.

7. The mix is complete when a chunk of it resists any efforts to pull it apart.

Mixing without a machine:

1. Form a pit by arranging 4 to 6 straw bales into a rectangle, then lay a plastic tarp over them.

2. Cover the tarp with a layer of clay that is about 3 to 4 inches thick.

3. Add some water and, using a shovel, hoe, or your hands, stir and push the clay around until it is uniformly wet.

4. Incrementally add sand to the clay and water mixture. Builders often mix with their feet by walking and stomping – or dancing – until the mix resembles a coarse, dense load of mud. Use caution when mixing with bare feet, as sharp rocks are known to lurk within.

5. Add straw and mix it into the mud. The consistency should thicken and become denser as the straw becomes mixed. Continue to add straw until it binds the mud and becomes thoroughly incorporated into the mix. Your mud will be sufficiently mixed when you can use the tarp to roll one big loaf that stays together.

6. If using, add the emulsified asphalt and mix thoroughly.

Once your mud is ready, form the bricks by doing the following:

1. Shovel the mud into the molds.

2. Use a rake or hoe to evenly spread out the mud within the molds and into every corner without creating air bubbles. When certain the mud is even and well-spread, use the back of a shovel or a trowel to level the mud and make the brick surface even with the top of the molds.

3. Allow the leveled bricks to sit, but do not wait too long, or they will stick to the molds and be difficult to remove without damaging them. This part of the process will probably take some trial and error, as the soil composition and water will determine the overall consistency of the bricks and influence their drying time.

4. When the brick is just dry enough to stand by itself, remove the mold. It is likely that because the adobe is still somewhat wet, the sides of the bricks may slightly bulge; this is all right because the brick will tighten as it dries.

5. Cover the bricks with newspapers once the molds are removed to stop the sun from drying them to the point of cracking or crumbling, or let the bricks dry under a tent. Both of these techniques slow the evaporation while the bricks dry. As with many aspects of alternative home building, this part of the process requires experimentation and trial and error.

6. Once the bricks are separated and covered, clean the molds to ready them for the next batch of adobe by hosing them down with water and scrubbing them with a stiff brush. If using small molds, it might be easier to completely immerse them in a tank of water and clean it with a brush.

7. Your adobe bricks will take several hours to dry enough to be handled and moved around without fear of damage. Move them to a designated area to continue drying.

After three or four days, stand each brick on its end to dry its remaining sides. This process allows air to reach every end of the brick. To ready the bricks for construction, they need a full six weeks of drying to reach a moisture content of about 4 percent. They can then be piled in preparation for building. Be careful not to put too many bricks in a pile to avoid breaking the bottom bricks. Protect the bricks from the effects of weather with plywood, plastic, or some kind of roof, and keep the bricks close to the construction site to avoid unnecessary labor.

Conduct several tests to ensure the bricks are reliable enough for building:

- Try to break a sun-cured brick with a bare hand. Turn it between the hands, rub at its edges, and place it on the ground and stand on it. It must sustain much more weight than that of a human being when included in the walls of a house, but if it does not break under these simple trials, the bricks are likely strong enough to meet the basic requirements for home construction.

- Or, drop a couple of bricks on their corners from a height of about four feet. The bricks should hold together with very little damage. If they shatter, the mix has too much sand.

Some new builders have a professional test the first batch of bricks to determine if the bricks are ready for construction. Professionals can test the amount of compression that bricks will withstand and measure their likelihood for rupturing. Find a lab nearby that can perform these tests on a few samples. Get a formal, written report on their assessment. After completing the bricks, return to the tester with a few more bricks to test again. If the bricks pass both the first and second test, the bricks have likely satisfied the requirements of the building codes in the state.

The process of creating these earthen bricks becomes easier and speedier with experience. The brick makers will also learn how to create stronger and more uniform bricks. On average, one person can eventually come to produce 100 to 150 bricks a day, depending on the size of bricks. Estimate the rough number of bricks in each wall section before making and laying them. Figure on about 160, 14-inch bricks or about 150, 16-inch bricks for every ten feet of wall. Builders plan on breakage, so prepare to have extra bricks by not allowing for window and door spaces in the walls.

Making adobe bricks is labor intensive, strenuous, and time-consuming work. However, plenty have successfully completed this task over the last few thousand years, and once completed, each of your adobe bricks will contain character, style, and a little bit of you. Those who have finished an adobe brick house will likely not attest to its ease of construction. On the other hand, you would be hard-pressed to find an owner-builder who opted for adobe bricks who is not proud of his or her work; you can probably even find some who would be happy to do it again.

Placing the Bricks

Creating adobe brick walls is a lot like creating a conventional brick wall except that the adobe bricks will be larger, and the mortar between the bricks will often not contain cement as a binding agent. It is important to keep each course of bricks level and the wall as a whole as close to perfectly vertical and plumb as possible. The tools used to do this are, of course, a level and a plumb line. Courses of bricks should be staggered so that a gap on one row is situated at the middle of the bricks above and below it.

Load-bearing adobe walls are uncommon; in most cases, the only weight the bricks will need to support is that of the bricks above them. Post-and-beam or stick framing typically carries the weight of the roof in adobe homes.

To begin laying the first course, spread a layer of mortar (about ¾-inch thick) on your foundation or stem walls. Place your first course of adobe bricks into the mortar and make sure that the course is straight and level; if a brick is not level, simply push it a little deeper into the mortar until it is. Once the first course of bricks is down, spread mortar on top of that course and lay the next course, all the while using your plumb line and level to keep your wall as straight and true as possible. In many cases, it is easiest to begin with the corners.

Doors and windows in adobe walls should always be bolstered with lintels. The window and door boxes described earlier in the "windows and doors" section should be appropriate for most adobe walls, but adding sills and lintels is a good idea if you want your walls to maintain good structural integrity.

Adobe brick walls are best constructed in the summer when temperatures will not dip below freezing. To create an adobe brick wall, do the following:

1. Begin by spreading a mortar on the foundation about ¾-inch thick. Adobe mortars vary in composition, but a 70 percent sand and 30 percent clay mix is a good place to start.

2. Place the bricks in courses, overlapping each successive layer over the joints in the course below by about 50 percent of each brick. In other words, the middle of the brick on the top course should overlap the joint of two bricks below.

3. Check frequently to ensure the bricks are level in each course. One method is to first lay two end bricks, nail string between them, and then use the string as a level guide throughout the course. Using a level is a good habit, too.

4. After completing each course, spread mortar in the joints to set the course. Use caution when applying mortar, as it will harden and be difficult to remove. Clean up extra mortar. Plastering over the adobe walls will help conceal extra mortar bits, which will be more obvious on non-plastered walls.

5. Place windows, doors, and lintels where desired and use half-bricks to complete every other course on the side of the window or door boxes. Make lintels 12-inches thick when they are above doors and windows that are up to about 8-feet long. Doors and windows that are more than 8 feet should have lintels that extend at least 6 inches beyond the opening.

CASE STUDY: ADOBE DOWN UNDER

Jai Goulding
Owner-builder

In 1989, Jai Goulding made up his mind to construct an adobe home on a five-acre lot outside Strathalbyn, South Australia. While he had the option of building a more conventional dwelling, Goulding was drawn to adobe for its affordability and efficiency. He readily admits that he had "very little" construction experience before he started the project; like many owner-builders before him and the many more to come, Jai found that constructing a home on his own with little previous experience is not only possible, but also fulfilling.

What began with a dream of self-sufficiency and affordable building eventually ended in a house that Goulding is proud to call home. He is happy to say that the experience and the house itself "has passed my expectations."

CASE STUDY: ADOBE DOWN UNDER

He started off with a little professional help flattening the site and forming the concrete footings of his foundation, but from there depended on his own two hands until it came time to install the large roof beams. The total construction time ended up being somewhere in the neighborhood of three years—the first of which was devoted almost solely to making bricks..

As for permits, Goulding made it plain that getting permission to build was not too difficult at all.

"I just had to submit the drawings and do a compression test on a brick," he said. "No professional signing was required."

Looking back on the project two decades later, Goulding is now sure that even the drawings for the house plan, which he opted to hire a professional to complete, could probably have been homemade.

Doing the work himself saved him on labor costs, but in the alternative construction world, it is not uncommon to get much more than you pay for.

"My house's energy efficiency is far superior to most conventional houses," Goulding said, though he readily admits the house is not without a bit of character. "It is marginally harder to heat up due to the high ceilings, and rain can damage the walls, so the walls need to be slurried [reinforced with a thin coat of plaster or mud] occasionally, but apart from that, it's no different to a normal house."

Not bad for a first-time builder.

"My house came home on budget and was very much less than a conventional house, as I did most of the labor required...," Goulding said. "I would not change anything; it went pretty smoothly."

One of the most enjoyable things about hearing the stories of owner-builders like Jai is the obvious connection that they feel for the place they live in. Unlike folks who purchase a house or hire a contractor to build one for them, Jai and his alternative building brethren know every nook and cranny of their homes with an intimacy that does more than scratch the surface. Perhaps this is why when you ask an owner-builder to pinpoint their favorite part of their house, there is often a bit of uncertainty: "That is hard to answer, as I am proud of the whole place... I guess the bathroom/laundry [room]."

CASE STUDY: ADOBE DOWN UNDER

By building with his own hands instead of with his wallet, Jai has the advantage of understanding his home's needs as well as its structure. When repairs pop up, he knows how to fix the problem himself; when maintenance is required, there is no need for a professional to step in. All in all, Jai's home is less of a box he paid to live in and more of an extension of his individuality and the land it is built on.

"[I] got exactly what I expected," he said.

Chapter 5

Rubber Tire Homes

The rubber tire that releases toxic fumes when burned and takes up space in a landfill is like gold for the right sustainable home builder. If so inclined, the sustainable home builder can take that tire, fill it with earth, and densely pack or ram the earth into the tire cavity with a sledgehammer. If that sustainable home builder wanted to give his or her latest creation a name, "earth-rammed tire" seems like it would fit the bill quite nicely.

Earth-rammed tires are heavy, cumbersome things, but when it comes to using recycled and naturally foraged materials to create a dwelling, few contraptions perform better. But it is impossible to speak of the many uses earth-rammed tires can be put to without mentioning one name and one movement. The Earthship movement, begun by Michael Reynolds, has made many homes with earth-rammed tires. Using these tires for foundational work, retaining walls and berms, and exterior walls, Reynolds and the Earthship movement have produced some of the most efficient, affordable, and striking dwellings the alternative construction world has ever known. His books and workshops have helped hundreds of people build their dream homes from little more than dirt and old tires; the debt all alternative builders owe to Reynolds and the Earthship movement is a large one, and anyone looking to build with earth-rammed tires would do well to take a deeper look at what this green-building founding father has to offer.

The Earthship building method is property of Reynolds, so if you want to construct and partake in the Earthship movement, be advised that purchasing his book or attending his workshop is essential. If you want to know how to build a house out of earth-rammed tires, however, there are a couple of things you can learn about the process here.

The process tends to be inexpensive because demand for used tires is low. In some cases, people will even pay you to remove used tires from their property.

For the earth-rammed tires to function well, the builder must pack the earth tightly for stability. Once they are well-prepared, the tires become excellent thermal mass structures, as the walls are about 2-feet, 8-inches thick. This method is typically used for a retaining wall that is built into the earth itself, with plaster serving an aesthetic function for hiding the tires.

Earthship designs use the tires for mass on the home's three sides. A wall of glass is used for solar gain on the south wall. The walls are U-shaped and built into earthen berms for structure and insulation, using the excellent thermal mass properties of earth-rammed tires. The walls are load bearing and can evenly distribute loads. While the thermal mass is high, houses in cooler climates require additional insulation for a more comfortable indoor temperature; even though the tire walls in the earthen berms have excellent thermal mass, the mass is not enough to maintain warmth inside during persistently cold weather.

Bottles and aluminum cans are also used in many Earthship designs. Aluminum cans are often used for inside wall insulation, glass bottles for light, and plastic bottles for insulation. Imagine a home that uses such a quantity of readily available recyclable materials that can serve such inexpensive, fascinating possibilities.

Earthships typically use the ecological concepts common to many sustainable homes, such as some type of water catchment system and use of grey water. Many such homes are built in the Southwest and have large, south-facing slanted glass windows to capture sun, which then can overheat the home in summer. However, it is certainly reasonable not to slant the glass, therefore reducing the solar gain.

Building Earthship homes in wetter climates can also be a challenge, as there are various obstacles to ensuring adequate protection against water. For example, the surface of the tires is irregular and therefore difficult to adequately waterproof. However, Earthship homes do exist in wetter climates, and for those interested in building one, it helps to go visit a few of the homes that were built to withstand such conditions. Research whether any exist in the state or climate region you are building in, take a road trip, and take notes on how this home's design deals with high moisture levels.

About Building with Used Tires

In some areas, there are building codes in place for earth-rammed tire homes. Even if these building codes do not exist where you plan to build, it is a good idea to take a look at what they have to say, as following these regulations leads to a stable and durable structure.

As earth-rammed tires are usually used to create load-bearing walls, installing a top plate (see the roofing section) is often a necessity. As earth-rammed tire walls are frequently also earth-sheltered walls, you can either create walls that are perfectly vertical (plumb), or design the walls to lean into the earth they will be sheltered by. Tire walls become substantially more complex when built more than eight courses high, or when a second story is desired. If you plan on two stories or tire walls higher than eight courses, hire a professional designer. Tires come in different sizes, but it is easier to use only one size tire in your construction. If you absolutely must use several different tire sizes, no tire should be larger than any tire appearing in a lower course. In other words, the bigger tires should be on the bottom of the wall.

Constructing Tire Walls

A typical tire wall is built with tires that have been tightly filled with dirt; this dirt is pounded again and again into the tire cavity with a sledgehammer — a very labor-intensive process. The reward for all this labor is a significant one: The wall you create by using this method will be exceptionally stable, strong, and be a very good insulator. Tire walls can be constructed without the arduous step of hammering earth into the tire cavity, but the method described here is the one that seems to be most commonly used.

Tire walls start with a solid foundation of either concrete or undisturbed, compacted subsoil. Tires are wide, so the foundation must be wider to provide sufficient stability. One of the main advantages of using compacted subsoil as a foundation material is that you can make it your foundation pretty wide with little expense. Sometimes, builders fill larger tires with concrete and use them as both a first course and a continuous foundation footing. Because the tire walls in many structures must be curved, a continuous, curved, perimeter footing will be used to create a suitable base for the heavy — often 300-plus pound — tires. Vertical rebar rods protruding from (or in the case of concrete, embedded in) the foundation can also be left exposed; the holes in the first course of tires can then be positioned around this exposed rebar to provide extra insurance against shifting for the first course of tires.

To create an earth-rammed tire:

1. Take an old tire and block the center hole on one side with a piece of cardboard. The cardboard keeps earth that is shoveled into the center hole from falling through before it can be rammed.

2. Shovel dirt into the center hole.

3. Using your hands, place the dirt into the cavity of the tire.

4. Once the cavity is sufficiently full, use a sledgehammer to compact the dirt into the cavity.

5. Repeat steps 3 and 4 until no more dirt can be pounded into the tire cavity. Depending on your soil composition, a single tire can hold up to three wheelbarrows worth of dirt.

To create an earth-rammed tire wall:

1. Once your tires are sufficiently rammed, place the first course directly onto the foundation; if protruding rebar was part of your foundation design (which is recommended), this will be a bit like a bulky, difficult form of ring toss!

2. The center holes in the first course of tires may be filled with many things. Often they are filled with concrete for stability, but they can also be filled with a sturdy material such as cob, which is cheaper. Concrete is generally the best idea for void fill on the first course and may be specifically required in some building codes.

3. Once the first course is in place and filled, the second course can be laid. Tires on the second course should have their centers directly above the point where the tires on the first course meet.

4. The center hole in non-first course tires has more leeway as far as their fill goes. Some use cob or light clay or mud; others use aluminum cans, and a few use plain, old dirt.

5. Continue building up courses in this manner until you reach your desired height. Walls using tires of uniform size should be no more than 6 courses high; if a taller wall is desired, the bottom row[s] of tires needs to be of a larger size.

6. Before the roof goes on, fix your top plate to the tire wall. This is often done by creating a continuous concrete bond beam or by anchoring a top plate of two staggered 2 by 12 wood boards to the tires with bent rebar.

CASE STUDY: WORKING WITH WHAT YOU HAVE GOT

Joe Dehne, Architect
SUJO Design Inc.
www.sujodesign.com
suz@sujodesign.com
joe@sujodesign.com

The idea of building a sustainable dwelling is an attractive one for many folks who are concerned about the environment. However, just as buying the most eco-friendly car on the market is a less sustainable move than simply keeping your current vehicle for as long as possible, constructing an environmentally friendly house is not always the greenest option for those who already own a house. Sometimes, building new is not as good a solution as upgrading what you have already.

"We started SUJO design in 2003," Suzanne Dehne said. "The main reason was to enable both of us equal time with our son. We provide architectural services to our clients, with the bulk of our projects being small residential remodels.

"Living in the city, our focus is more on remodeling existing spaces. It would be great to incorporate some of the more alternative building methods into urban settings, but we are not quite there yet. I would have to say I love the idea of re-use. Finding a second life for a product is just fabulous."

"We are striving to bring more green building methods and products into our work," she continued. "We hope to educate ourselves and our clients about what is available and, ultimately, include these features in our projects.

CASE STUDY: WORKING WITH WHAT YOU HAVE GOT

We were fortunate enough to be able to do some work to our own home, which allowed us to experiment with certain ideas and materials . . . so that we could really test them out and see for ourselves before making recommendations to clients."

Sustainable remodeling is an important trend in green building, and one of the great things about a remodel is that you get a new look — and sometimes a completely new space — without the high amount of labor most alternative structures take to construct.

"Our main goal is to provide our clients with an enjoyable experience as they venture through all the phases of their remodeling project," said Joe Dehne, Suzanne's husband, who is also an architect. "We design for minimal impact, energy efficiency, and healthy environments. We incorporate green building methods and materials where feasible. We strive to reuse, recycle, and reduce. The nature of architecture in and of itself is not very green, but we can help reduce our impact on the environment by designing smart."

The Dehnes enjoy the idea of creating a more efficient, sustainable living space by enhancing existing structures, but they are aware that sometimes creating a new house is the most attractive option. Here, their advice focuses not just on using better materials, but creating a better design, too.

"When possible, it's nice to incorporate passive solar elements," Joe said. "I have to agree with Suzanne: I like re-using as well, but more from an aesthetic level. There are simple design decisions that can be made that can be very cost effective [and may even be achieved] with no-added cost...The first one that comes to mind is building smaller. A family of four does not need to live in a 5,000-square foot house. A 2,000-square foot house will cost less to build, use fewer resources to build, and use less energy to run. Proper placement of windows for day lighting and passive solar are other key features."

The very popular concept of bigger and newer being better is something that the Dehnes are doing their best to change. By taking their advice, prospective owner-builders can certainly save some money — and will also be doing something good for the environment.

Chapter 6

Straw Bale Homes

For owner-builders in the United States – and likely plenty of other places, too – the notion of constructing a house of straw has been proclaimed a terrible idea for generations. Since childhood, we have been warned that neither straw nor stick is an acceptable construction material; we all know that those who disregard this warning will meet a single, big, bad fate: a wolf. Even still, the United States has a small but century-old history of constructing straw structures, and that history continues. In fact, the movement is gaining momentum today. From New England to the Midwest and to the Pacific Coast, many owner-builders are finding out first-hand that straw will not only keep out wolves but also be resistant to all kinds of huffing and puffing, as well.

When looking at various articles that talk about straw bale construction, you are bound to see one or two that tout the process as "fast." It is true that stacking the bales themselves goes very quickly when compared to creating walls of cordwood, rubber tires, or cob, but the amount of effort needed to make a straw-bale wall both sturdy and moisture resistant is significant. One time-saving advantage that straw bale construction does offer over some other alternative techniques: There is no significant aging or curing time involved in the process. While cordwood or adobe builders might devote months to constructing and drying bricks or aging wood, as long as your bales are dry, they are pretty much ready for installation.

provided by Carolyn Roberts

Straw bale homes are highly efficient buildings, particularly in hot, dry climates in which the cost of keeping homes cool tends to be high. Their R-value can be about R-30. The bales hold inside temperatures more steadily than materials in many other types of alternatively built homes. Straw bale homes have a natural cooling system that requires no electricity, making the home feel nearly air-conditioned even though the outside heat might exceed 100 degrees. This is attributed to both the high insulation value of the straw and to the interior thermal mass materials frequently used in straw bale construction, including the interior plaster on the bales. A skin of plaster coating protects the bales, stopping airflow but allowing moisture to penetrate, which greatly reduces the tendency of the bales to become moldy and rot.

Compared to building with conventional home-building materials, straw bales create a low ecological impact in that construction is fairly straightforward and chemicals are not required. Though some would argue that cereal grains tend to get sprayed with herbicides, this is not always the case. Construction involves stacking bales of densely packed straw tightly bound together. Post-and-beam frameworks in which straw bales are used to fill the space are the most common type of this structure, but straw bale walls can also be used as load-bearing elements. Each type has its advantages and disadvantages. For bales used as structural support, proper building techniques are required to ensure that the weight of the roof does not compress the bales. Although straw is the primary source in exterior wall construction, some estimate that straw-bale homes use only about 15 percent less wood than conventionally framed houses.

Straw bale homes can certainly be considered sustainable. Using local straw significantly reduces energy and transportation costs, and it keeps money in the local economy. Building homes out of straw that would otherwise be burned as waste is an excellent way to recycle this valuable resource; the U.S. contributes to the load of carbon dioxide in the atmosphere in part by burning 200 million tons of straw a year. Because of its one-year growth and harvest cycle, straw is a quickly renewable resource.

Straw is an excellent building material, but when using it, be aware that pests may be attracted to straw. Make sure you keep an eye out for infested bales during the building process and take bug problems seriously throughout the life of the house. While pests may pose a threat in a few cases, the biggest threat to straw bale homes is plain, ordinary water. The builder must know the threats that water poses and the different methods for preventing water damage to the bales. Thus, straw bale houses require much more protection in wetter climates than they do in dry. Climates with hard, driving rains can pose serious challenges, making straw bale better suited for drier climates.

However, well-built and carefully designed straw bale homes can and have held up remarkably well in wetter climates. As mentioned above, sealing the walls with some type of silicate paint helps prevent rot and also allows vapor permeability. Plastering interior and exterior walls might increase construction costs and lengthen construction time, but it will also greatly improve the home's longevity and durability. Straw bale homes built in Nebraska in the 1800s remain in use today and are beautiful, functional testaments to the durability of this building material.

About Straw Bale Building

If you grew up on a farm, you are may be already aware that there is a difference between straw and hay. True, they both come in big, yellow bales, but when it comes to building, one is far superior to the other. Hay is, of course, for horses. Cows and other animals also use it as a food source. Straw, on the other hand, is not generally fed to animals. Straw is the tough stalk left over after a grain like wheat has been harvested, and this toughness is what makes it the better building material. Hay can be found in just about any rural area; straw comes only from places where grain is being grown and harvested, which is probably why the oldest straw bale structures in the U.S. are found in the Midwest.

Because it cannot be found everywhere, some straw builders have had their straw shipped to them via truck. This practice will increase the cost of your project and decrease its greenness. If there is nothing else on your site to build with, however, shipping in straw is certainly a greener option than many others. If you live in an area where plenty of straw is available, you will have the advantage of being able to pick who supplies your bales. Good bales are tightly packed and bone dry; bales that have been stored inside are optimal. Straw bales can vary in size depending on how they are constructed, so it is a good idea to get all your bales from a single source whenever possible.

If you decide that straw bale construction is your preferred method, choosing a site and moisture channeling will be extremely important facets of your home's success. In the best scenario, you will have a site that is not only in a dry climate but also is in an elevated spot where water naturally sheds away from your structure. In other scenarios, you will need to make sure that your slab, stem walls, or piers are tall enough that moisture will come into contact with the bale walls as little as possible.

Straw bale homes should avoid running plumbing through the walls if possible, as running it through the floor avoids the destructive problems that come from moisture in straw bale. If plumbing does need to run through the walls, one option is to build a second wall in front of the bales to protect them from the plumbing that could potentially damage them.

When it comes to electricity, it is understandable that many would balk at the idea of running wire through a wall that is basically made of fodder. Of course, using exposed wire moldings can alleviate this fear, but surprisingly enough, the compact nature of straw bales makes running wire through them less of a hazard when compared to conventional construction with air gaps in the walls. Conduit systems placed within a straw bale wall can be metallic or non-metallic, but be sure to secure electrical boxes to wooden stakes that are driven deeply into the bales.

Building Straw Bale Walls

provided by Carolyn Roberts

Whether or not you want your straw bale walls to be load bearing, they need to be kept relatively high off the ground to protect them from moisture. This is often done by extending a stem wall up from the foundation. Some will say that a few inches above grade is high enough; in some areas, building code will determine the minimum height the bales must be above grade, and hence the minimum height of your stem walls. I would recommend constructing stem walls that err on the side of caution, particularly in areas that are not in a desert or almost desert climate. The concrete, block, or masonry used to create your stem wall will be more expensive than straw, but having your bales well out of moisture's way is something I believe is worth the extra money.

When constructing your stem walls, make sure you anchor some rebar in them. This is important, as your first course of bales will be pierced on this rebar to hold it in place and keep the bales from shifting. In some areas, code will determine the length of rebar that needs to be left exposed and the number of bars each bale on the first course will need to be impaled on. If there are no codes in your area, plan on having each first-course bale impaled on two pieces of rebar; also plan on extending the rebar enough that it goes all the way through the bales.

Straw bales take up more floor space than conventional walls, so be sure to incorporate this reduced floor space into the design of the house's floor plan. Permeable plasters such as earth plaster and lime plaster are good choices for straw bale exteriors, as their hygroscopic qualities allow moisture to exit the walls on dry days.

Because the danger from water and condensation is so great for straw bale homes, it is crucial to protect the straw bales from condensation at the bottom. A sill plate, which is a water-resistant horizontal structure added to the wall stem, will help ensure this. Go the extra mile to prevent water damage to the straw-bale home, and a strong home will prevail. Even after ensuring that the foundation situates the bales far enough off the ground, insulated sill plates attached to the stem wall further protect the bales by lifting them away from any condensation that might soak up through barriers below.

As far as the actual construction of the house, the way in which the bales are used will determine other design elements. Some people stack straw bales for infill between columns of timber and use it primarily for insulation, while others use the bales for the load-bearing task of supporting the roof. With water posing the most dangerous threat to a straw-bale home, the time it takes to cover and protect the straw is crucial. Load-bearing bales can much more easily become damaged by water during construction; because load-bearing bales create the structure and support the roof, they are set before the roof is placed. This order of events leaves the bales unsheltered and exposed to the elements. If it rains before the roof is put on, that can spell big trouble for unprotected bales. It is therefore important to make sure that load-bearing walls go up quickly, efficiently, and during a period where the chance of rain is as close to non-existent as possible. Creating a straw bale structure that is built within a post-and-beam frame allows you to put the roof on before the bales are stacked and also reduces issues with settling that often affect load-bearing straw walls.

The structure of a straw-bale home limits the builder's ability to adapt dimensions in certain situations. Bales that must be shortened to reduce length will need to be re-tied, or they will lose their structural integrity; trimming to reduce width can be done with a chainsaw. Reducing the height of a bale is not recommended, and it is better to design the wall height to equal the height of however many bales are intended for the wall. Straw pieces can be stuffed in gaps when using bales primarily for insulation, but this practice does not work for load-bearing, straw-bale walls.

For window frames, plan on designing them stronger for structural straw-bale walls rather than for infill straw bale. When a structure already exists, as for infill, the window frame — or buck — does not need to carry as much of a load, which consists mainly of the straw just above the window. Infill straw bale does not carry the weight of the roof. Bucks or window boxes provide added stability to the load-bearing straw bale house. They also provide a hard surface against which to stack the bales.

When placing a window, some builders try out different possibilities by placing bales in certain patterns to find the best location for the window before setting them. To keep a stable structure, place windows and doors at least one bale's length from wall corners.

Before stacking the bales, prepare the re-rod pins (bamboo, wood, or steel) for pinning the courses as they are stacked. Cut or sharpen the pins so they will anchor easily in the bales. The pins should be embedded in the foundation such that they protrude enough for impaling the bales for the first secure course. This process requires attention to detail, as it provides foundational support for the bales.

provided by Carolyn Roberts

Over time and under the weight of a roof, straw bales will compress and settle. To reduce this effect, load-bearing straw should be allowed to compress under the weight of the roof for several weeks before plaster is applied to their exterior. Plastic tarps that extend from the top course of bales down to the foundation are used during this compression period to protect the bales from moisture. Some builders also use thick, adjustable straps (like seat belts) to hasten this process; the belts are attached to both the bottom and the top of the wall at regular intervals and are then tightened to compress the bales.

To stack the bales for a load-bearing, straw-bale house, do the following:

- Be certain the foundation is complete and all the materials and help are ready to assemble the bales.

- Begin by placing the full-sized bales at the corners of walls and at door and window openings, stacking the bales toward the center of the wall. Overlap bales in corners for stability.

- To stack the second layer, course the stacks by staggering the bales on top of the bottom layer so that the middle of the bale lies above the seam below.

- Trim bales as necessary to fit as courses are laid. Chain saws can speed up the process, but use caution not to trim too much straw, altering the integrity of the structure. You can only trim the width of a bale with a chain saw (and this only works if you are laying the bales flat, with the straw ends facing the interior and exterior). In order to trim the length, you need to re-tie the bale using a "bale needle" or something similar. This is an essential tool for building with straw because you always need half-bales to keep the overlapping pattern in stacking the bales.

- After stacking bales fully around door frames, plumb the door box to be sure it is vertical, and attach the adjacent bales to the frame with dowels.

- Fasten window boxes to the second or third course of bales, depending on the height of the bales and window; fasten the surrounding bales to the boxes with dowels.

- Because the window and door boxes are made of wood and do not compress, allow some empty space above the boxes to avoid uneven settling as the roof compresses the surrounding top bales.

- As courses are stacked and positioned, pin them to the courses below by driving in wood, steel, or bamboo re-rod pins from above. Be sure to set the pins away from the bales' edges (the more centrally located the pins, the better).

- After the bales are secured and tied down, install the roof plate, which is typically either plywood or lumber, to spread the weight load of the roof. Some choose poured concrete, called a bond beam, because it bonds the top course of bales to the bottom of the poured beam, thus providing significant anchorage that prevents the roof being torn off in high winds.

- Follow one of these methods for tying down the roof to begin the compression phase:

1) Attach steel rods to steel fittings in the house's concrete foundation, placing the fittings and rods on one side of the bales.

2) Or, nail 2- by 2-inch steel mesh to a wooden wall plate that is nailed to the foundation, then nail the mesh to the roof plate at the top of the wall.

- To protect the bales from water damage while they are compressing, tie plastic tarps to the roof and secure them to the ground, ensuring proper coverage during the multi-week compression process.

- Before plastering, and while the roof is compressing the bales, smooth over any large bumps of straw that stick out from stacking or from compression. A weed eater, chain-saw, or grinder with a wire brush will work just fine.

As mentioned earlier, be sure the bales have completely settled before applying wall covering, such as earth plaster. Applying the plaster before the bales have completely settled will lead to cracks.

provided by Carolyn Roberts

CASE STUDY: WORKING WITH WHAT YOU HAVE GOT

Carolyn Roberts
Owner-builder
www.ahouseofstraw.com

Carolyn Roberts is not the only one who has been changed by the process of alternative home building. Her Web site, **www.ahouseofstraw.com** and her book, *A House of Straw: A Natural Building Odyssey*, are both doing their part to inspire others to take up straw bale building.

After seeing a TV show about Earthships earlier in the year, Roberts took the plunge into alternative construction in September 1999. Though she had no previous construction experience, she felt certain that she could build her own straw bale home on a bit of acreage outside of Tucson, Arizona. The area's dry, warm temperature was an ideal locale for a straw home, and her perseverance and attitude eventually brought her through the rewarding, yet strenuous, experience of construction.

The planning phase can be pretty complicated; as many owner-builders have discovered, when it comes to constructing an alternative structure, less is often more.

"The many little decisions about where to bring in the plumbing, where to put the outlets and the cooling ducts, whether to use gas or electric, and what type of roof were very complex," Roberts admitted. "I did my own design from seeing many homes when I was a realtor…I took it to a draftsman to be drawn into plans. Building smaller was what enabled this project to be a success."

Permits in place, Roberts began the process of making her home habitable. With the help of her two sons and a few friendly volunteers, she was able to prepare the site and build the foundation. Once the project was in full swing, she enlisted the help of consultant Jon Ruez. Even though she successfully completed a vast majority of the work with her own hands, having a good consultant can be as integral a part of building as the materials, themselves.

"I was very lucky to find Jon Ruez, who had years of experience in both straw bale and conventional construction," she said. "He lent me tools, showed me how to do the basics of the jobs (carpentry, drywall, framing doors, running electric wires, and tiling), found carpenters for difficult areas — like building the loft — and did difficult things like hooking up the electric wires, pex plumbing. [And he] helped carpenters with heavy tasks. He also told me what to do for each inspection and when to call the inspector…I would never attempt this without someone who knows construction as a consultant."

CASE STUDY: WORKING WITH WHAT YOU HAVE GOT

No one knows more than an owner-builder that constructing a house with your bare hands is not always a smooth process.

"Everything was more difficult and took longer than anticipated..." Roberts said. "There is no way a person can understand how much work it is to build a house until they have done it.

"I did this mostly by myself on nights and weekends while I was working full-time," she continued. "I moved in after one and a half years, but the finish work really wasn't completed for about four years."

That kind of extensive work can wear out even a seasoned professional, but Roberts sets an example for every prospective owner-builder: Not only can you do it, but if you take your time and keep your spirits up, you, too, can have a home you will be proud of. As to the finished home's performance, Carolyn could not be happier.

"The walls are extremely well-insulated, so the house is very slow to change temperature," she said.

"The sunroom is my winter heating; I have only an evaporative cooler that works well in the summer. I have no more maintenance needs than a conventional house...[My house is] far more beautiful and friendly than I imagined it would be."

Though Roberts is extremely happy with her finished house, there were a few things she'd do differently if she decides to build again.

"Have more money, more time, and know more about construction," she said. "Most of the building of a natural house is exactly the same as a conventional house. I had only studied straw bale, but I had a volunteer party and knowledgeable guide for that part. It was all the other days building the foundation, framing interior walls, windows, and plumbing that I really should have known more about."

CASE STUDY: LARGE-SCALE STRAW BUILDING

Habib John L. Gonzalez, Director
Sustainable Works
www.sustainableworks.ca
RR#1, S-4, C-12
Crescent Valley, B.C. V0G 1H0

Though building a single sustainable home is an excellent goal for the owner-builder, the people who call green building an occupation are trying to focus their efforts on increasing the scope of green building on a larger scale. Habib John L. Gonzalez has made it his goal to bring sustainable building concepts to the forefront of mainstream construction. Though changing people's minds is something that does not happen overnight, Gonzalez is certain that eventually alternative building and conventional building will, at the very least, meet in the middle.

"Plastered straw bale construction was introduced to me while working in Australia as a permaculture designer," Gonzalez said. "It provides a non-destructive use of otherwise waste straw in healthy, energy-efficient housing. Sustainable Works is an organization [that] works internationally in education, research, and building in the field of plastered straw bale construction."

Gonzalez and many of his like-minded contemporaries know that while making one house is a step in the right direction, ultimately getting commercial construction companies to accept green and alternative building practices will be a giant leap on the road to a safer, friendlier industry. As is the case in so many other situations, getting the mainstream to jump on the bandwagon happens through the action of individuals.

"The main areas of my work are incorporating the modern methods of plastered straw bale building into mainstream commercial construction, developing simple, efficient building designs and methods to meet local needs for affordable housing..." Gonzalez said. "All these methods I used on my own buildings before introducing them to colleagues and clients.

There are building professionals in British Columbia and Alberta whose interests and achievements in green building, particularly in plastered straw bale construction, make them important resources in their home regions. An important regional trend is blending passive and active solar heating systems with entry-level housing.

CASE STUDY: LARGE-SCALE STRAW BUILDING

While the general public has still not accepted straw as the terrific building material it is, Gonzalez's work is showing the world that a straw house is an option in more regions than you might expect:

"This area of Canada and the United States is an interesting patch to work," he said. "There are the tornado alleys of central Alberta, the deserts of the Badlands and Okanagan Valley, the heavy snow country of interior rainforests of southeastern British Columbia, and the earthquake zones of the coast and islands — plastered straw bale homes and commercial structures are in all these places."

As straw bale building continues to grow, technology and material availability is likely to grow with it. Thanks to the efforts of people like Gonzalez, the day may come when every town has a few straw builders, as well as a plumber and HVAC specialist. Who knows? We may even see the day when big box stores like Home Depot and Lowes have a straw aisle.

Chapter 7

Earthbag Homes

The term "earthbag" is pretty self-explanatory. You take a bag, fill it with dirt, and voilà: You have a building material that is inexpensive, long-lasting, and easily formed. Earthbags can be used in a variety of environments and can withstand many adverse situations. Though relatively new in the residential sector, the earthbag concept derives from the old sandbags historically used by the military for construction of barriers where temporary structures needed to be both easy to construct and literally bulletproof. This speaks to the difficulty of destroying this building material and reveals an important reason that alternative home builders have begun to hop on the earthbag bandwagon: This material uses earth and mostly natural materials for building, which means that it is not only eco-friendly, but affordable, too.

While burlap bags filled with sand were the materials that triggered this movement, experimentation has brought a shift in the material that is used for both the filler and the bags. Although burlap bags remain an option, they do not last as well as other materials. Polypropylene bags are now used with more frequency than burlap. If you plan early and do not mind doing a little exploration, you might be able to find polypropylene bags that once served another use and are scheduled to be scrapped. For instance, misprinted rice and grain bags are often available in bulk at a discount.

About Earthbag Building

Earth alone can be used to fill the bags; however, many are finding that other earth-based materials work just as well. Everything from volcanic rock to adobe soil can be used to fill the bags, with the volcanic rock being a superb, moisture-resistant, and insulating material. It is important to note that the bags themselves do not hold up well when exposed to direct sunlight for extended periods of time, so it is necessary to cover them with a finish material. Many alternative builders prefer a natural plaster or papercrete for this task.

Building Earthbag Walls

The two types of bags most often used for earthbag building are burlap and polypropylene. Less durable and often more expensive, burlap should not be used when the fill is only sand. Polypropylene bags are popular and easy to find, but are more prone to sun damage and absolutely must be sealed with papercrete or earth plaster as quickly as possible. While these two types of bags are the most common, any type of bag that will last a long time is worth considering, as long as you have easy access to a large supply.

When it comes to filling the bags, there are a wide variety of soils that will be fitting, so use whatever is most abundant at your particular site. As with other earth-based building materials, a mix of sand and clay works well, although be aware that soil with a high-organic content, like topsoil, will likely settle over time. Also consider more than just soil — volcanic rock, crushed shells, or gravel work well. Should volcanic rock be local to your site, it is excellent for earthbag construction because it is light and insulates quite well. Even rice hulls — another natural material with good insulation value — have recently been used to fill earthbags.

Some soil contents settle, while others do not. Volcanic rock tends to settle somewhat, but sand does not. Lighter materials, such as peat or woodchips, settle more than do heavier materials. When using a fill that is light, it is recommended that bags be packed well and tamped down to reduce settling. Although some builders opt for longer bags— which become long tubes upon being filled — such earth-stuffed bags may be difficult to handle.

Follow these steps to fill and place the earthbags:

- Moisten the soil to make it stable. Dry soil is not stable because it shifts, and cement is sometimes added to bottom bags that must carry an extremely heavy load, or to bags that will be used in forming arches. The first course or two of bags can be filled with gravel to prevent moisture from going up into the wall.

- Fold in the bags' corners to avoid creating protrusions, thus preventing large bumps in the wall.

- Shovel the soil into the bags close to the site.

- Once the earthbags are filled to your liking, they must be properly sealed. Some builders recommend leaving just enough unstuffed material at the top of the bag to be folded under as the bag is placed horizontally on the course. Others opt for a fuller bag that is then either stapled or sewn shut.

- Bags are ready to be laid once they are filled. Lay your first course of bags onto the foundation.

- Once a full course of bags is placed, thoroughly tamp them down to ensure their stability and evenness. Do not tamp until a full course is placed, as tamping beforehand will create unevenness throughout a course.

- Use a long-handled tool with a heavy, flat bottom to tamp until the bag no longer gives. A simple tamper can be made by placing a long stick in wet cement that has been poured into an empty plastic container.

- Consider a course thoroughly tamped when the bags no longer give and a ringing tone is heard.

- Stabilize each course by placing between them a material such as barbed wire to create friction. The friction will reduce slippage.

- Once the courses are placed and tamped, the bags need to be covered as quickly as possible to avoid sun damage.

- For window and door openings, rough frame before lying courses for each particular wall so you know where to create the necessary gaps. Self-supporting windows can be made with bags by forming arches, circles, or triangles pointing upward.

Chapter 8

Cob Homes

In many situations, the term "cob" describes not a building material, but rather the inedible item left over after an ear of corn is consumed. While it is probably possible to construct a dwelling from such leftovers — if you try hard enough, it is possible to build a house out of just about anything — in alternative building, cob means something very different, indeed.

Much like the previously discussed adobe, cob is an earth-based product that has been used for thousands of years, dating back to settlements 10,000 years ago. People sometimes consider cob structures to be both the oldest type of construction in human history, as well as the least complex. Historians and archaeologists have found examples of cob buildings all over the globe, from the Middle East to West Africa to the American Southwest, and even to the British Isles, where the term itself originates. Cob is an English word that means "a lump," or "loaf." In construction terms, a cob structure describes a building made from earth that does not use bricks, mortar, or wood. This sustainable material's growing popularity is partly due to its malleability, making it ideal for creating uniquely shaped homes. Cob is fireproof; according to its biggest fans, cob is also a lot of fun to work with.

Like adobe, cob is made by combining clay, sand, and straw. The major difference between these two types of earth-based building materials is not the ingredients used to make them but the way in which they are used in the construction process. While adobe is shaped into uniform-sized blocks and left to dry naturally, cob is applied directly to the foundation, usually by hand or with basic tools. Cob is what builders often choose when they want to sculpt the home because of its malleability and lack of pre-defined form. Another difference between cob and adobe is that cob tends to be less expensive; it is not currently as trendy as adobe, nor does it require as much supporting material.

About Cob Construction

While building with adobe can be an easy process to envision, cob construction is even simpler. The building process is done by hand and usually without design plans that are as specific as for adobe. Cob building does not require mortar joints, which makes it less prone to earthquake damage. Cob walls are usually covered with lime or earthen plaster, or are whitewashed for protection against the elements. This outer protection significantly reduces weather-related damage, and the thickness of the walls helps regulate the inside temperature and provides a high thermal mass value. In addition to earthquake resistance, cob houses are also pest resistant and fire resistant; if properly designed to keep moisture at bay, these structures can last a long time.

People who want to do all building themselves often choose cob. But this is still no small task. While building with cob is simple in concept, the construction process is very labor intensive. Though other alternative building techniques require the use of some specialized equipment, many modern cob builders create their cob by hand (or feet). Although it is possible to use larger mixing equipment to create the cob material, this practice tends to minimize the earth-conscious efforts of the builders and therefore defeats some of the reasoning for building with cob in the first place.

Like adobe, cob absorbs heat; as mentioned above, the thermal properties, or thermal mass, help maintain a comfortable temperature year-round. Its insulation value is not that high – about R-0.5 per inch – but it will increase with the thickness of the wall. A three-foot-thick wall could possibly attain an R-value of 15, depending on the locale. Site specifics do make a difference. A cob home in a climate in which the temperature remains at or below freezing for extended periods can become quite cold without thorough insulation. In fact, it is probably advisable not to build with cob in such climates, as the insulation has to be extremely efficient to counter the cold air that the thick cob walls will draw in from the outside.

Cob also shares adobe's soundproofing qualities, but the most significant factor these materials have in common is a small ecological impact. Cob lacks the toxins of conventional home materials, and the fact that few materials or equipment need to be transported to the building site means that its negative effect on the environment is near zero. Because the earthen materials help regulate the home's humidity, asthma and other breathing issues in people are also typically improved. Like adobe, cob is a hygroscopic material; cob walls will have the ability to absorb and release water in response to the humidity of the air around it, resulting in more stable interior humidity levels. Though cob's hygroscopic nature is beneficial in this respect, the fluctuating moisture levels of the material can create problems with mold. To avoid mold problems, it is important to ensure that the straw used in the building process is clean and pristine; using straw that has already been compromised by rot or decay in cob walls increases the likelihood of mold formation over time.

The financial expenditure for constructing a cob exterior is extremely low, especially when compared to a conventionally built home. Without having to buy or import a significant amount of any man-made substance (milled wood, metal, or cement), much elbow grease – more than money – is necessary to erect a cob wall. Because the majority of the home's main material is found in the subsoil near the site, the only other possible expense for the wall materials is the purchase of straw, sand, or clay from nearby sellers, if none is available on the property. In many cases, the material for the walls of a cob home might cost the builder about a few hundred dollars in total. Exterior walls are only one financial cost in building a home, however; claims that cob houses are often constructed for only a few hundred dollars are prevalent, but misleading.

Labor is by far the biggest ingredient in a successfully constructed cob shell. If you are paying for labor, you will find that your costs rise even faster than your walls. This kind of project is, of course, less expensive if you can depend on friends and family to provide some of the grunt work. The necessity of manual mixing and quick application of cob to the foundation while it is at the right consistency means that optimal cob construction requires a crew rather than an individual; while it is probably possible to build a cob wall on your own, the process will progress more smoothly with a crew of at least three or four.

By encouraging neighbors, family, and friends to participate in the building process, your project can quickly become a community event. While construction crews offer experience, they are also expensive; on the flipside, men, women, and even children can participate in building a cob home without having any previous construction experience. A family of four could construct their home together, and each member 5 years old and up could have a job he or she could do successfully. Many have found that there are more than monetary benefits associated with this route – quite a few relationships have been forged and or solidified during cob construction, so do not be afraid to invite the in-laws.

Building Cob Walls

Cob home builders must also be aware of the damaging effect that water can cause to the structure. Although a coat of lime or whitewash normally protects the outside material from a certain amount of weather, structural additions can lessen certain types of water damage. Placing a cob home on top a hill with no natural protection potentially exposes the home to a greater amount of water damage from wind-blown rain. Placing a cob home in a ditch where water is likely to pool is also a poor choice; if possible, opt for a site that is protected from high winds but also situated so that water will naturally flow away from the structure. Eaves that hang far out from the roof will be necessary, and attached gutters to ensure water does not settle at the base of the home can be helpful, as well. Cob walls should be at least 6 inches above the surrounding ground to protect them from moisture – you may want to raise them even higher in wetter areas. Many find that constructing stem walls from stone masonry is an aesthetically pleasant way to lift cob away from ground moisture.

Increase the stability and lifespan of cob walls by tapering them at the top. By creating a wide, strong base that narrows into a thinner top, the stability of the wall increases. The thicker the wall, the more insulating and stable the house will become. Make the thickness somewhere between 12 to 24 inches wide for the base of a cob wall, using thicker bases for taller walls. Build for stability. Especially in earthquake-prone areas, do not build more than two stories. Though the material is especially good at withstanding shaking at most heights, the lower each wall's center, the less likely it is to sustain damage. Consequently, design the home with a maximum of two stories, particularly near areas close to fault lines and seismic zones.

Designing curved walls in a cob structure is another way to create stability and resistance to the elements. Because typically designed square or rectangular rooms are the most vulnerable to cracks, and tend to collapse along their edges and corners, a curved wall avoids these pitfalls by adding strength and steadiness.

When planning openings in cob walls, the narrower the distance between a door and window, the less stable the walls will become. Extremely wide doorways or windows will also lessen the wall's strength. Try to limit the size of the openings, and position windows far enough apart that the cob between them will be thick and sturdy. One benefit of building with this material is that it is possible to construct the entire building, then go back and cut the windows and any interior recesses after the cob is placed.

After determining the locations and sizes of all the openings, it is helpful to consider both practical and aesthetic reasons for choosing certain types of windows and doors. Some cob home builders just use fixed glass within a windowsill, which is considered the easiest method of getting natural light into the house. Others find that they want serviceable windows that they can open and close to fully take advantage of light, heat, and air.

One of the most important aspects of protecting the cob home is making sure the interior and exterior are properly plastered. Opt for an earth plaster that breathes so that moisture can escape back into the air, and make sure that your walls are dry before applying the plaster coat.

When installing plumbing in a cob house, the builders can simply place the cob around the pipes while building the walls. Pipes can also be added after cob has been completed by first wetting and then cutting a deep enough channel into the cob; later, simply lay a few inches of freshly mixed cob over the pipes to cover them up. Electrical conduit can also be embedded within cob walls, but as is the case with many other types of alternatively constructed exteriors, items embedded in the walls can be rather difficult to access at a later time.

Making Cob

Be sure to perform a soil test, as described in the section on adobe, for determining the soil composition on your building site. Soil that has a high silt content is typically not good for cob. Clay and sand make a good cob mix, and there is certainly a possibility that you will have to purchase one or another to supplement your existing subsoil. Soil of a high organic content, like topsoil, will settle over time and is typically not considered a good addition to cob mixes; however, it is unlikely that a bit of topsoil that slips in here or there will create a large impact on the finished structure.

Creating cob is like creating adobe, except you skip the process of creating individual bricks, and you use a bit more straw. Like adobe, cob can be made with the help of a mixing machine, or you can let your hands and feet, along with a few hoes and shovels, do the work. If using a machine, make your first test batch of cob by following these steps:

- First, add 7 to 8 inches of water in the bottom of the mixer.

- Begin shoveling in clay on top of the water and turn on the mixer.

- Mix the ingredients, adding water and clay as needed to create a stiff mud. Work out all the lumps and clusters. This mixing will soak the clay in preparation for the next step: adding sand.

- Add sand to the clay already in the mix. At this point, some builders prefer to mix the soaked clay and sand by hand – or by foot – instead of by machine because the mixture becomes dense and heavy.

- If using an emulsifier, slowly add the agent according to the directions.

- Let the mixer work for a couple of minutes until the solution is thoroughly blended, then add the straw. A ratio of about 1:5 in straw to mud is an approximate goal, which will become evident when the mixer slightly slows down. If it does not, slowly add more straw until the mixer reduces its speed.

- The mix is complete when a chunk resists any efforts to pull it apart.

If building *without* the aid of a machine, do the following:

- Form a rectangular pit by arranging 4-6 straw bales around an open center. Lay a plastic tarp over the bales. Then add the clay in a layer that is 3-4 inches thick.

- Add some water and using a shovel, hoe, or your hands, stir and push the clay around until it is thoroughly mixed and uniformly wet.

- Then incrementally add equal parts sand to the clay, and mix until it resembles a coarse, dense load of mud.

- Next, add straw and mix it into the mud. The consistency should thicken and become denser as the straw becomes mixed. Continue to add straw until it binds the mud and is incorporated. It is well-mixed when the tarp can be picked up to roll the mix into a loaf that retains its form.

- When mixing the mud, rocks or lumps of clay left in the mix can provide extra aggregate once the walls harden.

- Form the mud into manageable-sized lumps or "cobs" to be used as soon as possible in the home's structure.

Creating a cob wall is merely a matter of adding lumps upon lumps until you have a continuous, monolithic shell. Instead of making courses of individual components like bricks, bales, tires, or logs, when a cob wall is created, it is a single entity; every lump of cob you add will be worked and smoothed until it is no longer a discernible lump but an extension of the wall itself. Building cob walls is a far more free form process than building walls with other materials and, depending on the builder, the final structure can be as conventionally square or as innovatively askew as you like, as long as the wall's integrity is sufficient enough to be stable.

Once your foundation and stem walls are in place, you can begin constructing your cob wall. This entails simply three steps:

1. Place a lump of cob in position.
2. Place the next lump next to or on top of the original lump and work the material until it creates a seamless mass.
3. Continue this process until the desired dimensions are achieved, taking care to keep the vertical elements of the wall plumb.

Compressing each course of cob is a fourth step that is far more important for load-bearing cob walls than it is for cob used as infill between post-and-beam framing elements. Compressing cob (or "thwacking") is typically accomplished by beating each course with a 2x4, cricket bat, or similar wooden instrument while it is still wet. This will create denser, more stable courses and a wall that is better able to handle the weight of a roof.

The number of courses you lay in a day has some bearing on how the finished wall will look and, in some cases, function. Laying at least one course a day is recommended, as skipping a few days in between courses allows the cob already in place to dry too much and will make integrating fresh cob more difficult. This is even more of an issue in very hot, dry areas and is a likely reason why adobe is a more popular choice for building in such climates, even though the materials function very similarly once the structure is complete. If you lay many courses in a single day, the large mass of wet cob is more likely to bow and bend before it dries. Each team of builders will have to figure out on their own how many courses make a good day's work.

As you build your walls higher and wider, keeping them straight and vertical will be a challenge. It is certainly handy to use a plumb line and horizontal guide strings as a references during the process, particularly when trying to create rectangular structures. Cob is a very forgiving material in that it can be cut and mended both during the building process and after it is dry. Though the free form nature of cob building is very appealing in several ways, having a plan before you actually begin creating cob walls is a wise choice that will help you create a structure that is not only beautiful, but stable, as well.

CASE STUDY: CREATING HEALTHY LIVING SPACES IN A HEALTHY MANNER

Andrew Morrison, Owner and Managing Member
Straw Bale Innovations
Straw Bale Workshops
A.C. Morrison Construction
A.C. Morrison Consulting
www.StrawBale.com
www.StrawBaleWorkshops.com
www.CoachingBuilders.com
www.AskMrGreenBuilder.com
7803 Sterling Creek Road
Jacksonville, Oregon 97530

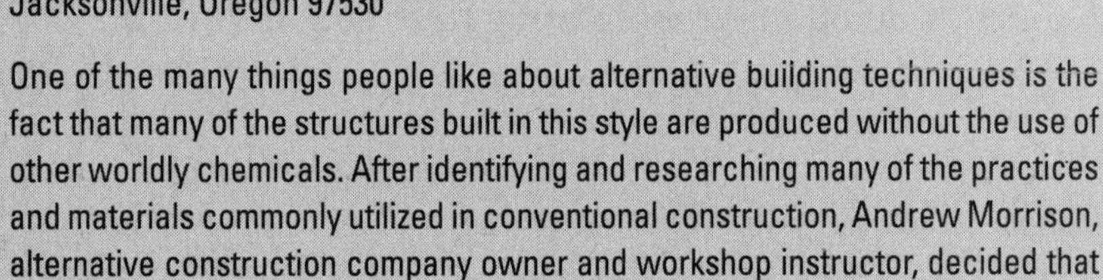
One of the many things people like about alternative building techniques is the fact that many of the structures built in this style are produced without the use of other worldly chemicals. After identifying and researching many of the practices and materials commonly utilized in conventional construction, Andrew Morrison, alternative construction company owner and workshop instructor, decided that the time had come to try a different approach.

CASE STUDY: CREATING HEALTHY LIVING SPACES IN A HEALTHY MANNER

"I have always been dedicated to caring for my community and my planet, and efficient, healthy construction was simply an extension of this long-held belief," Morrison said. "I first started researching alternative home building options back in the late 1990s. At the time, I was working as a geotechnical drill manager, sampling soils and bedrock conditions for new construction. This exposed me to the fact that large developments produce massive amounts of waste. There was literally enough 'waste' for me to build a home addition with the discarded material. This wastefulness unnerved me, and so I started investigating alternative home building methods."

Wastefulness was not the only part of the construction equation that Morrison found disturbing. Anyone who takes a look at the materials that have been a part of the modern home over the past 100 years or so is bound to turn up some pretty scary stuff. Asbestos insulation and lead paint might be the best known of conventional construction's more recent mishaps, but questionable materials are still being used every day in buildings all over the country. One of alternative construction's main goals is to bring our dwellings back to a more natural standard, and Morrison works on doing this whenever the situation allows.

"In order to be truly healthy and 'green,' we need to consider [eliminating] products like zero Volatile Organic Compounds (VOC) glues and paints [in favor of] natural flooring like cork and bamboo, and whole-house heat recovery ventilation systems that provide healthy indoor air quality while maintaining the majority of the heat in the home," he said. "...I take a holistic approach to building."

"I currently own several companies, all of which focus on creating healthy homes and individuals through 'alternative' methods," Morrison continued. "I have dedicated my professional life to the expansion of these views and hope to one day remove the term 'alternative' from these practices and replace it with "efficient" or some other term that moves these practices out from the fringes of society and into the mainstream. I teach workshops, am the host of a video series called "The Straw Bale Minute," and have authored articles and e-books on green design and construction. In addition to my work life, I also incorporate green practices into my personal life, driving a bio-diesel auto, buying locally, and producing most of what we eat...I have made minimizing my footprint on the planet a central part of my life...It is my passion to learn as much as I can about alternative products so I can provide new and exciting options to my clients.

CASE STUDY: CREATING HEALTHY LIVING SPACES IN A HEALTHY MANNER

By teaching people how to build healthy homes, we hope to inspire them to invite others into the journey."

Morrison is a champion for all those willing to go the extra mile when it comes to sustainable living, but he knows as well as anyone that this is not always the easiest path to tread. Creating a sustainable dwelling can, of course, be physically demanding, but many people find out the hard way that this process can sometimes present a mental strain, as well. In order to prepare his clients for the other-than-physical stresses that come from building a house, Morrison makes sure that his workshops and consultations include advice on mental preparedness.

"I am available for personal coaching as well as planning, design, and construction consulting," Morrison said. "You may be surprised by how emotionally difficult building a house or having a contractor build one can be...I can help with the emotional support and personal growth work, as well as the physical work. They are equally important to the healthy construction of your home."

Chapter 9

Sod, Rammed Earth, and Compressed Earth Homes

provided by Michael Blaha

Sod, rammed earth (or pisé de terre), and compressed earth can each be used to create a permanent dwelling. Though these are all earth-based materials, they are each very different from one another in both form and installation. Sod is basically a section of grass and topsoil that is cut from the ground. Rammed earth is a moist mixture of dirt, water, and a stabilizer (often lime or cement) that is poured in a form and then stomped, tamped, and rammed until its size is considerably reduced, forming a monolithic, compressed earth wall. Compressed earth block is rammed earth that is formed into large bricks rather than installed in place. Each of these materials will have formidable thermal mass and insulate properties. Sod and rammed earth are also, literally, dirt cheap.

About Sod Construction

provided by Michael Blaha

Sod construction has taken place for centuries; in fact, some sod homes that are more than a century old are still standing today. They are solid structures but require the right elements to work. Vikings built sod houses; Americans also built sod homes as they settled the West. Though one might initially balk at the idea of building a house from dirt and grass, sod's availability and tested longevity make it an excellent choice in certain situations. Just as early Midwesterners saw plenty of grass and not too many trees in the frontier landscape, people today who live in an area where wood is scarce can consider sod as an option for a new dwelling.

The difference between sod and other earth-based materials is that sod is earth that includes grass. In contrast, adobe — the most popular earth-based material — is used in areas where grass is not as common and the earth itself is the most easily accessible building material. In places that are wet and grass grows in abundance, owner-builders should not dismiss sod as a feasible building material. The earth in these areas may be softer and less fit for constructing adobe or cob, but the dense roots of grass that hold the earth together creates a solid building material that is both affordable and plentiful.

Sod does not tend to sustain rain damage well and therefore requires an outer wall covering, such as stucco. It suffers erosion and is generally difficult to cut. Yet it maintains fairly regular temperatures, is fireproof, and can easily be found in areas where grass is abundant – hopefully within the locale where the home is to be built – making construction of a sod home very inexpensive.

Some sod houses that have been in existence for a hundred years or more, such as the Addison Sod House in Nebraska, are built into the earth with only one side taking the elements. This sort of structural support greatly reduces the costs of building the walls, increases the thermal mass, and ultimately lowers energy consumption. Like any other building material – alternative or conventional – sod has both benefits and drawbacks; accentuating sod's positive aspects is typically a matter of good design combined with finding a good place to build.

Excellent regions for sod homes – or soddies – are those where the sod is plentiful. Locating the right kind of sod for the most efficient alternative home building is key. Look for areas where the earth is soft but is held together tightly by the roots of readily growing grass. Moisture content is also important in cutting sod for construction. Find land that has a moisture content that facilitates easy cutting but is not overly soggy and wet. The primary location for finding sod is in the Midwest. It is best harvested in the fall, after the roots have penetrated the soil deeply. Some sod builders recommend using prairie grasses over others, such as fescue, which is less dense. Other good choices include big and little blue stem, wiregrass, and wheat grass. Use grass with densely packed roots, as it is stronger than grass with looser roots. Make sure that the sod is not only suitable, but plentiful as well; estimates on older sod houses gauged about an acre of sod for a 12- by 14-foot house.

Historically, sod houses have made an excellent dwelling both for temporary and permanent uses, but have always fared better as smaller structures. In a small soddy, the walls are typically stable enough to support themselves without the aid of an interior frame; a few posts are often used, however, to support the weight of the roof. Creating a sod house is probably not the best idea for those who want a house with all the modern conveniences, but as a hunting cabin or getaway, soddies can function well, be constructed cheaply, and have an attractive (albeit rustic) appearance that will fit into a natural setting.

Constructing Sod Walls

provided by Michael Blaha

Building a good sod wall starts by cutting sod. This process is relatively straightforward, but its practice can be grueling. To facilitate efficiency, some sod builders make their own sod cutters. As you might imagine, the market for manufactured hand sod-cutting devices has not yet reached proportions that make such items readily available; however, sod cutters made as tractor attachments, which cut long patches of earth in designated thicknesses and widths do exist. Used mostly on sod farms, these devices can be handy, but the sod patches they produce still generally need to be cut into individual pieces for building purposes.

Sod should be removed after mowing the grass to about 4 inches tall. Cut sod into strips about 4 inches thick, minimum, and about 12 inches wide and 2 feet long. Move the sod to the home site either on pallets with a forklift or by pulling it on a trailer with a tractor. It is probably not necessary to construct an elaborate foundation for your soddy; the significant weight and width of sod walls is generally enough to keep the structure in place. But as an earth-based material, it is a good idea to set sod walls above the surrounding ground on stem walls, or on a gravel trench to keep moisture away from the sod itself.

Typically, the sod bricks are placed grass-side-down.

1. Lay the first and second courses of sod pieces lengthwise around the perimeter of the intended house, making sure to stagger the joints between courses. Your walls will be 2 feet thick, so in these first two courses you will have to lay two sod pieces side-by-side to create this width.

2. For the third course, and every third course hereafter, lay your sod pieces widthwise. This will hold the middle joint between the first two courses in place and add stability to the structure.

3. Continue building up your walls as described until you reach the desired height, making accommodations for door and window frame boxes. Window and door boxes in sod walls should be heavy and strong; they should also be fitted with dowels or keys (see the section on windows and doors) to help keep them in place.

When you get to top course (which should not be too high, as one story is fine for sod structures), it is time to attach the roof. This can be done in several ways—including anchoring a wood top plate into the sod walls to which the roof can be fixed—but one of the easiest is to simply lay heavy beams across the expanse of the roof that are supported both by the sod walls and posts on both sides, either inside or outside of the house. The weight of the beams should be sufficient enough to not be easily moved or blown away, but it is a good idea to use anchoring devices like spikes and nails sunk into both the wood beams and the sod walls to keep the beams in place.

About Building with Rammed Earth

Rammed earth is a mixture of sand, clay, gravel, and occasionally a stabilizer, such as Portland cement or lime, which is then tamped within a sturdy frame to compress the mixture into a solid structure. Post-compression, the walls' frames are removed, allowing the walls to cure in the warm sun.

Rammed earth walls can take two years to fully cure, but they become stronger during this time. The wall essentially becomes akin to solid rock, but because it is earth and remains porous, rammed earth walls are typically sealed to prevent water damage. Rammed earth structures are more easily maintained in drier climates, but they can be built just about anywhere. Also, they last a long time in comparison to conventionally built homes. As far as costs, professionals estimate that building a rammed-earth house comes in at no more than two-thirds of the cost of a conventional frame house.

In the United States, rammed earth gained popularity in the 1930s when the government, through the U.S. Department of Agriculture (USDA), developed a community of such housing. The government built the houses and sold them with small plots of land for affordable prices to low-income families. After World War II, however, the popularity and use of conventional building practices skyrocketed with the decreasing cost of conventional materials.

As with other earth-based building materials, rammed earth uses local subsoil; this keeps the topsoil intact, leaves most transportation and fuel costs behind, and creates a structure that uses less overall energy than most conventionally built houses. The walls transfer moisture to keep the inside naturally comfortable; the thermal mass is high to reduce heating and cooling costs; and the material is naturally insect, fire, sound, and mold resistant.

Many people mistake adobe and rammed earth as the same building material, partly because of their similar appearance. The difference lies in the way that the mixture is formed to make the home. The rammed-earth builder typically, but not always, uses machinery to compress the earth to create flat surfaces that offer clean lines and thick walls. The use of the machinery trades human energy for petroleum-based energy, increasing the embodied energy in building the home. While rammed-earth construction can be and has been done without the aid of machinery, the use of powered tamping devices speeds the process dramatically. The decision to use tamping machines or not is based on how green the builder wants to be, how much labor people can provide, and how much money the builder is willing and able to spend.

Rammed earth is strikingly similar to CEB, the difference being that rammed earth uses the same process to create a large mass of compressed earth, rather than individual bricks for building. As previously mentioned in this book, building with dirt-filled recycled tires is another method of creating rammed-earth walls. Both traditional rammed earth and tire construction can produce a green, sustainably built home, but the tire method is probably better for owner/builders who want to use as little mechanized energy as possible during construction.

For plumbing a rammed-earth home, it is best to place plumbing systems and their supply lines outside the wall under cabinets designed for them. Access will obviously be easier, and there is no risk of crushing the lines when compacting the earth. Water supply lines that absolutely must be placed in rammed-earth walls should be encased in a heavy-duty plastic sleeve to prevent moisture from coming into contact with the wall, and the system pressurized and observed during the compacting stage. Waste lines that must be placed in the wall should be insulated and the earth carefully compacted around it. If using power tampers, do not tamp directly on the pipe.

Electrical wiring, outlets, and switches can be mounted on the surface of rammed earth walls after the compaction is complete, but can also be placed in the walls if the proper steps are taken before compaction. By putting a "place saver" (in laymen's terms: a wooden block) in a designated space, pulling the block out after compaction will give you a quick and easy space to install outlet boxes or switches. Electrical wiring inside a heavy conduit can be put in place before compaction, but because tamping wires that have electricity run through them can be a touchy task, it is advised to work with somebody who has experience with the process.

Constructing Rammed-Earth (Pisé de Terre) Walls

Labor intensive and time consuming, but eco-friendly and long-lasting, rammed-earth walls have a look all their own. Though a few will end up coating these walls with stucco, plaster, or earth plaster, the appearance of a clear-coated, rammed earth wall with its layers and striations is truly something to behold.

A rammed earth wall is only as good as the form it is rammed into, so take particular care to make sure that the forms you build are as solid and stable as possible. The use of sturdy wood panels and braces is a must. As forms increase in height, stability becomes more difficult to create. Rammed-earth walls are generally at least a foot thick, but can certainly be made thicker to provide more stability.

For the soil, the consistency and moisture level is important; moisture helps the soil bond, and a lack of moisture creates weak walls. Too much moisture causes the walls to shrink as they dry and, in severe cases, this can even cause cracks.

Prepare the soil mix in the following way:

- As mentioned for other methods of soil mixing, prepare the soil as close to the work site as possible to reduce unnecessary transporting.
- Measure 30 percent clay and 70 percent sand and mix with shovels.
- After the soil is mixed, add water while continuing to mix, until the soil achieves a consistency of moisture that allows good compaction. Be sure the mix is not so wet that it will shrink significantly as it dries. Follow this tried-and-true test: Form a small ball of soil with both hands such that it holds its form after opening the hands. If it crumbles, the

mixture needs more water. If the ball holds its shape, drop it onto a hard surface, and if it crumbles into loose bits, it is appropriate for construction. If the ball holds its shape when dropped, there is too much water in the mix; let it dry a bit before using.

- When the soil's consistency is sufficient, shovel the moist soil mix about 6 inches deep into the forms. Using hand rammers or pneumatic tampers, work the mix until the first layer is dense and hard. If tamping by hand, hold the tamper a foot or so in the air and let it drop onto the soil. One or two people usually stand inside or around the forms as they tamp.

- Many builders attest to doneness when tamping makes a ringing sound, not a dull or hollow sound, as loose dirt makes. Be sure the corners are well-tamped, and watch that the soil is not over-tamped. Stop when the ringing sound is evident.

- Continue adding 6 inches of soil and tamping until dense and hard, filling and tamping until the soil is just below the top of the form.

- Immediately disassemble the form from around the molds and reset them to layer a course on top of the bottom course.

- Begin a wall from a corner. To make corners, simply ram wall sections against each other. However, ramming can create problems for the outside corners, and care must be taken to keep the walls square and straight. Builders sometimes avoid this by not having wall corners meet directly but by extending one wall beyond the other.

- When designing the home, forming windows and doors will be much easier if the measurements coincide with the shape of the panels used to form the walls.

- Place a volume displacement box (VDB) in the space where the window or door will eventually be set. The VDB must have the exact dimensions of the window or door, be square, and be sturdy enough to resist any deflection that can occur during tamping. Simply set the end boards against the openings for the sides.

- Some use actual bond beams for lintels, as they must be able to support very heavy loads above. A lintel, when the bond beam cannot support the structure, can be either solid wood, concrete, or steel.

- As with adobe, make lintels 12 inches thick when they are above doors and windows that are up to about 8 feet long. Extend lintels at least 6 inches beyond the opening over doors and windows that are more than 8 feet.

- Place rebar in the footings in earthquake-prone areas.

- Lay bond beams across the top bricks for a bearing plate for the roof to evenly distribute the weight.

The roof of a rammed-earth house — as for any earthen structure — needs to protect the walls from rain. Before placing the roof, set the bond beam and add sill plates, which are typically made of wood, and fastened to the beams, posts, and rafters. Roofs can be flat in dry climates, and the coverings can be made of any number of materials, including wood shingle, slate, or sod.

About Compressed Earth Blocks

provided by Michael Blaha

Compressed Earth Blocks (CEBs) are made by taking soil and compressing it to form large, sturdy bricks. The best soil to use in constructing CEBs will be subsoil with a content of up to 30 percent clay and 70 percent sand or silt, but sturdy blocks can be produced using soil of various compositions. Building with CEBs provides a very strong structure. The blocks can have a compression strength up to 2,000-pounds per square inch (psi), with the average somewhere around 1,200 psi. Because the material for CEBs is taken from subsoil, the rich, fertile topsoil left over can be used for gardening or creating a living roof.

Several factors can determine the appropriateness of building with CEBs, including the availability of labor, and the timeframe for construction. While most of the alternative building methods described in this book require a substantial amount of labor, making the actual blocks of compressed earth is an additional process that will add time to the overall project length.

However, unlike adobe bricks, which must spend a long time drying out, after the CEBs are made, they are ready to be used in assembling a wall. Because CEBs have much lower moisture content than adobe or cob, the house's placement and location is less of a concern. The inner humidity of the home is also less affected by the weather or season, as the walls absorb and release moisture according to the environment.

One advantage that CEBs have over other sustainable building forms is their uniform shape. The blocks are made with a press that creates the same size block over and over again and generates less waste than many other methods. The uniform shape of CEBs makes the use of pre-made building components, such as windows, doors, and roofing, much more feasible. Although an experienced builder can certainly help the process, it is unnecessary for the first-time builder to acquire many new skills when using CEBs, making this a preferred home building material choice for people who want to create their homes with their own hands.

If desired, the CEBs can be shaped to interlock with each other. Such a technique reduces the need for mortar and other binding materials that can be wasteful, unnatural, and sometimes environmentally toxic. As with other earth-based buildings, CEB buildings have a high thermal mass, are sound resistant, fire resistant, and insect resistant, as well as mold resistant. Unfortunately, like other earth-based materials, CEB's insulation properties are quite low, typically less than R-1.

Home construction with compressed earth blocks is a fairly recent development in alternative home building. It is believed to have been first used in the 1950s in South America and became easier with the development of the Cinva Ram, a manually powered device that creates one block at a time using a lever system. The 1970s and 1980s saw a rise in its popularity, particularly in the southwestern United States. Several countries in Europe are beginning to use the process, with some creating building standards, indicating its emergence into the mainstream construction practice.

Builders in developing countries use this material because of its availability, lack of required material transportation, and the minimal amount of professional knowledge and experience necessary for using it to build solid homes. Many nonprofit organizations that specialize in home building regularly use CEBs. Habitat for Humanity, the Peace Corps, and the U.S. Agency for International Development all have begun to use CEBs in their construction efforts.

Building CEB Walls

provided by Michael Blaha

After a suitable foundation is formed, you can begin building CEB walls. The first step in creating a CEB wall is, of course, creating many CEBs. There are several types of CEB machines available, and the creation of blocks will depend on the type of machine you are using. Prices for CEB presses range dramatically; in a quick search, I found a few manual models that cost between $2,000 and $4,000 — and several automated models that were about $10,000. It may also be possible to purchase a used CEB press at a discount or even rent one if you live in an area where this type of construction is common. It is important to either follow the directions that come with the machine, find operating instructions online, or have someone on-site for the first few runs to demonstrate how the machine is properly used. Once you have the process down, making blocks should be simple.

provided by Michael Blaha

After about 24 hours, the blocks should be ready to be handled and stacked several layers high to continue making room for the next batch. If a large crew is working on the project, it is possible to lay the bricks onto a wall after it has dried for a day, either stacked dry on top of one another or set together with mortar. With a group of people, it is entirely possible to build the shell of a small, simple house with one day for block production and one day for stacking. If in a hurry, various mechanized tools, such as soil crushers, sifters, or mixers, will speed the process. With mechanized tools, organizing the team and streamlining the process simplifies and speeds up the repetition necessary to produce CEBs.

Lay the bricks by doing the following:

- As with adobe, lay the first bricks on a layer of mortar (using a wet slurry of the actual CEB material is fine), which will bond the bricks in place.

- Stack them in courses masonry-style – as with typical kiln-fired bricks – with the middle of the bricks in the course above positioned over the joint of the bricks in the course below.

- Be sure the wall height-to-thickness ratio is stable. Wall height and thickness may be subject to building code in some areas.

- Once the desired height is reached, set a reinforced concrete bond beam on top of the walls to connect the bricks to each other and to provide a base upon which the roof will be attached.

- Windows and doors for CEB structures can be either purchased from a manufacturer or constructed in a box as described in the windows and doors section of this book. If using purchased windows and doors, make sure to follow the manufacturer's guidelines for installation; if building window boxes, lintels and sills should be used to increase stability.

provided by Michael Blaha

Chapter 10

Aluminum Can and Glass Bottle Homes

Though recycling has gone a long way in reducing the number of cans and bottles that wind up in our landfills, the idea of taking such materials and putting them directly to use in construction can be even more environmentally appealing. Over the past few decades, alternative builders have been experimenting with empty drink containers and turning them into houses; the results have been mixed, to say the least.

There have been cases where bottles and cans have been used as a supplementary material in forming walls and windows in other types of construction, but the trial-and-error process of finding a good, stable, efficient way to use such materials as the primary building blocks of a dwelling is far from over. To say that one cannot create a home from empty bottles or cans would be a lie; it has been done. To say that such structures are a feasible means to an efficient end, however, would be a little misleading. If you already have a collection of several thousand empty drink containers, then attempting to build with these can be done. However, if you are looking to build efficiently, effectively, and affordably, there is a good chance that there are materials in your area that are better suited for construction than a beer can is.

That being said, as the human race continues to become more environmentally conscious and concerned, we will find all sorts of interesting building materials and techniques that will be used to reduce waste. Empty drink containers certainly have a glimmer of promise in this respect. If you are considering a structure that is more for play than for permanent residence, perhaps you will have some luck with these materials, but as for laying out an efficient and time-tested way to construct an actual dwelling from cans or bottles, one simply has not yet been established.

About Can and Bottle Construction

Cans and glass bottle have their place in alternative construction, but rarely is it the case that any one of these materials has been successfully used to construct a permanent dwelling.

At the Heineken® headquarters in the Netherlands, you can view a wall made entirely of bottles called WOBOs (short for world bottles). These are glass beer bottles that were specifically designed to be used as a building material, shaped like bricks with a short neck on the top to drink or pour from. Their design also allows one WOBO to fit relatively snugly to another WOBO — kind of like beer bottles that resemble LEGO® building blocks. Heineken did not produce too many of these WOBOs (the idea did not really catch on), and there are only two permanent structures built from them, both of which are owned by Heineken.

Glass bottles have been used successfully in constructing a permanent dwelling when used in conjunction with logs in cordwood masonry. Completely replacing the logs with glass bottles has been done, and the structures produced have been interesting; however, not too much research is available on how well these structures perform. The practice of laying bottles along side logs in a cordwood wall is common. If the bottles are cut in half, and the two bottom halves are put together to form a glass cylinder, they can be covered with metal flashing and placed in the wall just as you would place a log to create a small, circular, "window" the size of the bottle's bottom. As an alternative to cutting, you can also tape the bottles together with duct tape and wrap them in metal flashing to create a less transparent window. You can also use the same method but replace the bottles with glass jars, creating a transparent "log" without cutting any glass. When several clear bottle or jars are placed adjacent to each other in the wall, you will have what is basically an inoperable window.

If you visit Houston, Texas, you can see a house made entirely of beer cans, but this house took a long time to make, and the process was rather tedious. It is uncertain how well aluminum cans perform in rainy weather or as an insulated material, but it is unlikely that beer cans are as nearly as effective a building material as the others outlined in this book.

Aluminum is one of the most readily recyclable materials; the process of recycling aluminum cans and using the metal in production again is more efficient than mining aluminum ore. Though cans have been used as a siding material, combined with papercrete to form non-load-bearing walls, and to fill voids in rammed-earth tire walls, other materials that are not so readily and efficiently recycled can do each of these jobs. Not only will these other materials work just as well as aluminum cans – or, in some cases, much better – many will be a better choice for anyone looking to save energy on a large scale, as these materials – unlike aluminum – would otherwise go to waste.

If you happen to have been saving aluminum cans for the past several years and have thousands of them stockpiled and in good condition, there is certainly no harm in experimenting with what you can do with them. However, as many places offer 5- or 10-cent redemptions for each can, your best bet for getting the most out of your collection is probably turning those cans into cash. Not only will you pocket a little profit, you will also be helping to reduce energy consumption in a very real way by allowing new aluminum products to be produced without the high energy job of mining.

Chapter 11

Sustainable Accents for Alternative Home Builders

The shell of a home is not the only area where sustainability can be a factor. Plenty of energy is expended in the process of manufacturing finished home components like cabinetry and flooring; even raw materials like lumber and natural stone require a lot of energy to get to showroom condition and to be transported from their place of origin to your home. Many of the components found in your average house contain materials that are far from natural, but there are often natural or sustainably produced alternatives to just about every thing. By choosing interior components and accents that are sustainably produced, locally manufactured, or salvaged from another site, you can greatly reduce the imprint your home makes on the environment.

CASE STUDY: ONE SURPRISINGLY SUSTAINABLE ACCENT

Jim Bryant, Managing Partner
Vintage Copper
www.vintagecopper.com
info@vintagecopper.com
3215 N. 448 Road
Salina, Oklahoma 74365

Though it may not always seem like the case, green building is not all straw and mud. While certain materials seem to scream sustainability, others that we have been familiar with all our lives turn out to be surprisingly eco-friendly options. Case in point: copper. Though nothing about copper's attractive, unique appearance is as overtly environmental as, say, mud bricks, its use in construction poses many benefits to the conscious builder.

Jim Bryant, managing partner at Vintage Copper in Salina, Oklahoma, knows that copper does not have to oxidize to make it a green building material.

"We build copper building material for the exterior of the home," Bryant said. "Most, if not all, of our products are on or near the roof. For example, we build cupolas, filials, vents, chimney caps, and guttering. I became interested in these products after first being exposed to them because they are truly unique from other products available out there. This was something that I could take pride in.

"No one wastes copper material," he continued. "The motive for this is a financial one; it is too expensive to waste this material. But the result is one that is beneficial to the environment; nearly all scrap copper gets recycled on the job. Additionally, these are long-lasting products that are designed to withstand the test of time, further offering an environmentally healthy option. Our products are thought by many to be luxury products, but most will last longer than their owners. Our vents are passive and work without power. The concepts behind our products are not new. Most came from a time before electricity, and they are designed for the person interested in lasting value."

It is true that copper materials are often more expensive than their competition, but that expense pays off in a big way when you consider how long copper lasts. Long-lasting materials lead to fewer repairs and replacements; this, in turn, leads to less wasted energy. By choosing a high quality material — that is naturally occurring, to boot — the environmentally conscious builder reaps benefits from all sides.

CASE STUDY: ONE SURPRISINGLY SUSTAINABLE ACCENT

While other materials come and go, classic, natural materials that perform well are likely to be a part of constructions — both conventional and alternative — for a long time to come.

"My partner started Vintage Copper 40 years ago," Bryant said. "He is one of the leading copper craftsman who has trained numerous other people in the industry. He has worked on projects as small as individual homes and as large as restoration of the Statue of Liberty. I joined him eight years ago and have seen continuous developments in the industry since that time."

Adapting old materials to perform in new ways is a large part of what green building is all about, and Bryant is as excited as anyone about innovation and new uses for existing technology.

"I think that experimentation in home building is always a good thing," he said. "Much of what is proposed in the midst of trying out alternative home building turns out to be inefficient or impractical. However, it becomes a stepping stone for developing better products. I believe in a natural selection process with these building materials. Many of the new home design systems will fail; some will not. Those that pass the test will probably enhance our lifestyles. Any new procedure, technology, or design that we come up with that saves energy, decreases upkeep, and increases comfort and security are along the lines of what I think we would all like to see."

Bamboo

The resilient bamboo plant was used for homes more than 5,000 years ago. Its tensile strength has made it an optimal building material for bridges in China, and it has been used to build boats and even early airplanes. While the tendency is to call bamboo a type of wood, it is literally a very tall, strong, and flexible grass — so strong, in fact, that its compressive strength is about two times that of concrete and about on par with steel.

Most bamboo species are grown in southeast Asia and Latin America. The types grown in more temperate or high-mountain climates are called running bamboo, which has a long underground rhizome, from which bamboo is reproduced. The other type is climbing bamboo because the rhizomes are short, and the bamboo that grows from them is contained in a smaller area and has thicker-walled culms (stalks) that have a larger diameter than the running type. Though some bamboo species are solid, most are hollow inside.

The bamboo that is used as a flooring material or a structural element grows very quickly. Both lumber and bamboo are generally harvested at maturity; however, the rate at which trees and bamboo reach maturity is remarkably different. Hardwood trees like oak and walnut might take 50 years or more to reach maturity. The bamboo used to make flooring, on the other hand, can be harvested after a mere five or six years.

Today, bamboo is used frequently throughout Asia as a structural building material. This concept has yet to catch on in the United States. However, bamboo is rapidly becoming a popular flooring and accent material in North America and throughout the world, partly because of its extraordinary versatility and its rapid rate of growth. It can be used to strengthen the structures of homes built primarily with other materials and can also serve as a strong and flexible roofing material. Compared to timber, bamboo is not only much stronger and more sustainable, but has a soft feel and unique appearance that is simply not offered by any other material.

Cork

The bark from the cork oak tree is the ubiquitous stopper in wine and champagne bottles. But cork had different uses thousands of years ago. For example, the Romans made shoes from the spongy material, and it began showing up in flooring in the late 1800s as an underlayment. Alternative home builders use cork today for flooring, as it is soft, sound-absorbing, resistant to mold and water damage, and renewable. Cork also insulates, reducing energy use and bills. As an insulator, cork can be used as an underlayment below hardwood or bamboo flooring. Millions of cell walls inside the cork create an excellent insulator as well as shock absorber.

Advocates for using new cork focus on how the material is harvested — not from cutting down the tree, but from harvesting its bark, which immediately re-grows. It is harvested in nine-year cycles from a one species of trees that grow mostly in Portugal and in Spain. Because of this short cycle of growth, it is replenished fairly quickly and does not require the killing of a tree. Cork trees often live from 150 to 200 years; if you do the math, one cork tree can easily be harvested 15 or 20 times.

In building, cork is used in several places. Its acoustic-absorbing properties make it a great material to use on walls and ceilings; it can also be used for gaskets and seals — think of the champagne bottle). Cork's most popular use in the building world, however, is as flooring. It is durable, and its appearance is often considered both interesting and pleasing. Cork is also a resilient material, making it pleasant to walk on. Cork naturally contains a waxy substance called "suberin," which makes it moisture resistant and protects it from rot.

Reclaimed Lumber

Reclaimed lumber is exactly what it sounds like: lumber taken from another source and used in a new installation. In many cases, reclaimed lumber is taken from old barns, houses, and other buildings that are set to be demolished. When taken care of, wood can last a very long time, and when a structure comes down, it is better to reuse the salvageable wood it contained rather than have it end up in a landfill. Even wood that has been somewhat neglected can often be brought back to usable condition, and the look of this aged wood is like nothing else available.

While reclaimed lumber obviously does not require the cutting down of trees to produce, the process of making very old lumber suitable for new installation can be just as energy-intensive as producing lumber from trees. Due to this, it is often the case that reclaimed lumber is even more expensive than conventional lumber. When it comes to the sustainability of reclaimed lumber, some suggest that the large amount of energy involved in making old wood suitable for new installations detracts from the material's greenness. However, you can avoid both the high cost and less sustainable aspects of preparing old lumber by salvaging it yourself. If a building in your area is coming down, ask if you can have a look inside it. The beautiful wood paneling and flooring of old office buildings, and even the posts, beams, and rafters in old barns can often look great in a rustic setting (such as an owner-built house) without much additional attention.

Salvaged Materials and Components

Wood is not the only product that can be salvaged from older structures. Light fixtures, sinks, bathtubs, doorknobs, tile, mirrors, windows, and many other components are pulled from structures set for demolition, saved by salvage companies, and made available for purchase. You may even be able to cut out the middle man by finding structures that will soon be torn down or gutted and asking the demolition contractors for access to some of the materials that would be going to a landfill. The dumpsters around buildings that are being renovated can be gold mines for the right-minded builder, as well. Often, many components that are thrown away are in good condition — though usually in need of a good scrubbing — and are being chucked simply to make room for a newer, flashier installation.

Recycled brick, also called salvaged brick, is a material that is catching on in the mainstream. The look of old brick might be weathered, but its performance is much the same as new brick; when an old brick building comes down, companies that specialize in collecting the bricks strip them of any clinging mortar and resell them. If the building is old enough, the bricks may be handmade rather than machine-produced, offering an even more unique look. Sometimes when old roads are renovated, brick-like pavers that were used before asphalt became the standard are discovered and can be salvaged. These pavers are extremely strong and durable and look great when used as a walkway, driveway, or patio.

Using salvaged materials in your home is more eco-friendly than using new materials, but the character of these materials is often just as enticing a reason to invest in them. Finding companies that deal in salvaged materials can be as simple as looking in the phone book or typing "salvage" into your favorite search engine. Be aware that some salvaged components may be valuable, and companies who deal with them may charge more than you expect for some pieces. Finding demolition and renovation sites – and their particularly fruitful dumpsters – to scour for materials can be a bit tricky. When getting ready to build, keep your eyes open as you drive from place to place, and pay attention to construction and renovation areas. Talk to the workers at these sites and ask if they would mind you going through their refuse; if they are paying a landfill by weight, there is a good chance they will be happy to let you take whatever you like.

Wool Carpet

Wool carpet is considered one of the most comfortable additions to the naturally built home, as it provides a soft, colorful alternative to the traditionally hard surfaces generally found here, like wood, concrete, and tile. When compared to synthetic carpet, wool is the more expensive option; however, for some alternative home builders, the luxuriousness that wool carpet offers coupled with the fact that it is a natural material ends up being worth the added cost. Wall-to-wall carpet can be a liability in high-traffic areas as it holds onto tracked-in dirt and debris more efficiently than hard-surface flooring. Due to this dirt-attracting tendency and its high cost, most find that the addition of wool carpet to their floor plan works best in specific areas where comfort is most appreciated; bedrooms, where people want a soft first-landing of the day, are a perfect example of where wool carpet would be best.

When compared to many synthetic carpet fibers, wool is naturally more resistant to dirt, stains, and wear. It is flame-retardant and does not support bacterial growth, plus, it helps floors have insulation, where heat loss can be high in colder climates. Wool carpet is nontoxic when naturally dyed; some manufacturers offer 100-percent natural wool carpet that does not contain the glues or chemicals you might find in other carpet products. Aside from being completely biodegradable, such flooring would not have the typical VOCs that reduce indoor air quality, either. Though different types of natural fibers are becoming more common in alternative homes, wool remains an excellent and long wearing soft insulator for floors that many eco-minded homeowners will appreciate.

A Completely Recycled Home

Being a green builder does not necessarily mean that you must be an alternative builder. There are many conventional products on the market now formed from recycled content; even the most traditionally constructed house can be made using a high ratio of recycled materials. Carpet, roofing, siding, flooring, and even concrete can be produced from recycled materials – you just have to keep these materials in mind when shopping around.

Theoretically, a home made entirely of recycled materials is possible. Using recycled gravel and concrete for the foundation; reclaimed lumber for framing and flooring; salvaged windows, doors, and fixtures throughout the house; metal roofing made from recycled content; and even reclaimed plumbing and wiring, you could create a traditionally constructed – and traditional-looking – house. When you combine such recycled components with alternative building techniques, the home you create can be unique, energy efficient, and have very little impact on the environment.

Knowing that a recycled home is possible is a far cry from actually constructing a dwelling completely out of recycled materials. In most cases, the biggest obstacle will not be the construction, but the legwork involved in finding all of the components you need. If this is your goal, start planning early. Some conventional constructions are planned years before the first screw is ever turned; if you want your home to be composed of recycled materials, be aware that you might have to devote a similar amount of time gathering materials and information.

CASE STUDY: GREEN HOME BUYING?

James Rogers, Real Estate Agent
Green Key Real Estate
www.greenkeyrealestate.com
James@greenkeyrealestate.com
444 Eighth Street
San Francisco, California 94103

James Rogers originally began studying sustainability as an economics undergraduate during a course he took while studying in Madrid, Spain. After seeing how far ahead in this issue that Europe was compared to the United States, his awareness and interest were piqued, and he realized the importance of incorporating sustainability into every aspect of our lives. When he learned how much of U.S. greenhouse gas emissions come from the building industry, he realized that this was one of the most important areas for change.

CASE STUDY: GREEN HOME BUYING?

Rogers is involved on the real estate transaction side of alternative home building, and his influence will have its greatest effect with home remodeling because the majority of home buyers remodel at least one room after home purchase.

"If I can guide my buyers to use alternative home building techniques in these remodels, making the homes more energy-efficient using environmentally friendly building methods, I have done my job," Rogers said.

Rogers work as a real estate agent with Green Key Real Estate, which is the first and only green real estate company in the city. Its focus is on incorporating environmentally responsible and socially just principles and practices into all aspects of its business. Green Key Real Estate runs an office waste minimization program, powers the offices with wind energy, respects tenants' and home owners' rights and responsibilities, and donates a portion of profits to promote green building practices. The goal is to empower, enrich, and educate clients toward green-minded buying, selling, and remodeling of their homes. Green Key Real Estate works to make sure that environmentally responsible and safe decisions are made with these purchases, connecting buyers to green service providers and products to make a community more sustainable.

"Many realtors can walk you through a house and show you the kitchen and the backyard, but how many can tell you about easy upgrades to make your home environmentally friendly?" Rogers asked. "As a Green Realtor, I can help you in all of your home buying and selling needs, while adding a level of service and knowledge that you really can't find anywhere else at the current time. I can help with all of the traditional aspects of the sale, but can also add information about indoor air quality, sustainability, energy efficiency, and resource conservation."

Rogers is a Certified Green Building Professional through Build It Green, a Bay Area organization committed to increasing awareness of green building and also an EcoBroker, which means that he has been trained in the environmental aspects of homes and how they relate to the buying and selling process.

Conclusion

Building your own home is a challenge of the highest order; it can be fun to watch your plan turn into a dwelling, but no method described in this book is what anyone would call easy. The size and scope of your project will have a lot to do with how difficult the process ends up being, and my advice to any first-time builder is to start small. If you have ever wanted a little cabin in the woods, or even a tool shed on your property, this is probably a good setting to get some practice before jumping into constructing a permanent residence. Having a small project to start off with will give you a better idea of what a larger project will entail and help you solidify your technique at the same time. Plus, if you make a few mistakes, the consequences will be far less drastic than if those mistakes were made in the construction of your home.

Most owner-builders will continue to do work on their homes well after they have started living in them. Problems arise, components are replaced or upgraded, appearances are enhanced, and amenities are added throughout the life of these structures. Thankfully, anyone who builds his or her own house is typically capable of working on it without the aid or the expense of a professional; at the very least, he or she will be good at differentiating the do-it-yourself projects from those that will require professional attention.

It is easy to become discouraged during an alternative building project, particularly if you did not realize how much work goes into such a job. Many of us have seen houses go up in half an hour on TV, and likely even more of us have heard of communities coming together and building a house or raising a barn in no time. But this will not be your experience. With plenty of help, you may end up with a project that goes from beginning to end in a few months. Good help, however, is hard to come by. For most of us, the process will be a long one, taking place during our spare time – and often with only one or two people lending a hand. Do not fret. The hours are long and often difficult, but in the end, the rewards are monumental.

Just as important as the time you spend on building your home is the time you spend away from the project. Owner-builders typically begin their projects with conviction and determination, but the sacrifice of free time – and the aching joints and muscles – eventually takes its toll. Do not get discouraged; it is important to try to keep the same pace and attitude you had in the planning of your project throughout its entirety. Once the project is in full swing and you have a good idea of what kind of work can be accomplished in a day (or a week, or a month), make certain that you plan some time away from the site to relax and enjoy yourself. Find a few good stopping points, and make sure to treat yourself during these times to some R&R and a few activities that you may have put on hold to work on your house.

Alternative building, in all its forms, is a practice in trial and error and learning as you go. Remember, even if you feel a little lost at first, by the time this project is finished, you will be an expert – at least, an expert on your own home. You will have an intimate knowledge of your home's construction that those who hire out their work will never have, and when (if) problems arise in the future, you will be thankful for this experience.

Creating an owner-built home is a difficult path, but it can be tread by anyone who really wants to follow it. Chalk your setbacks up to a lack of experience, and celebrate your achievements wherever they arise. In the end, you will not only be a happy homeowner, but you will be a happy and knowledgeable alternative home builder, too.

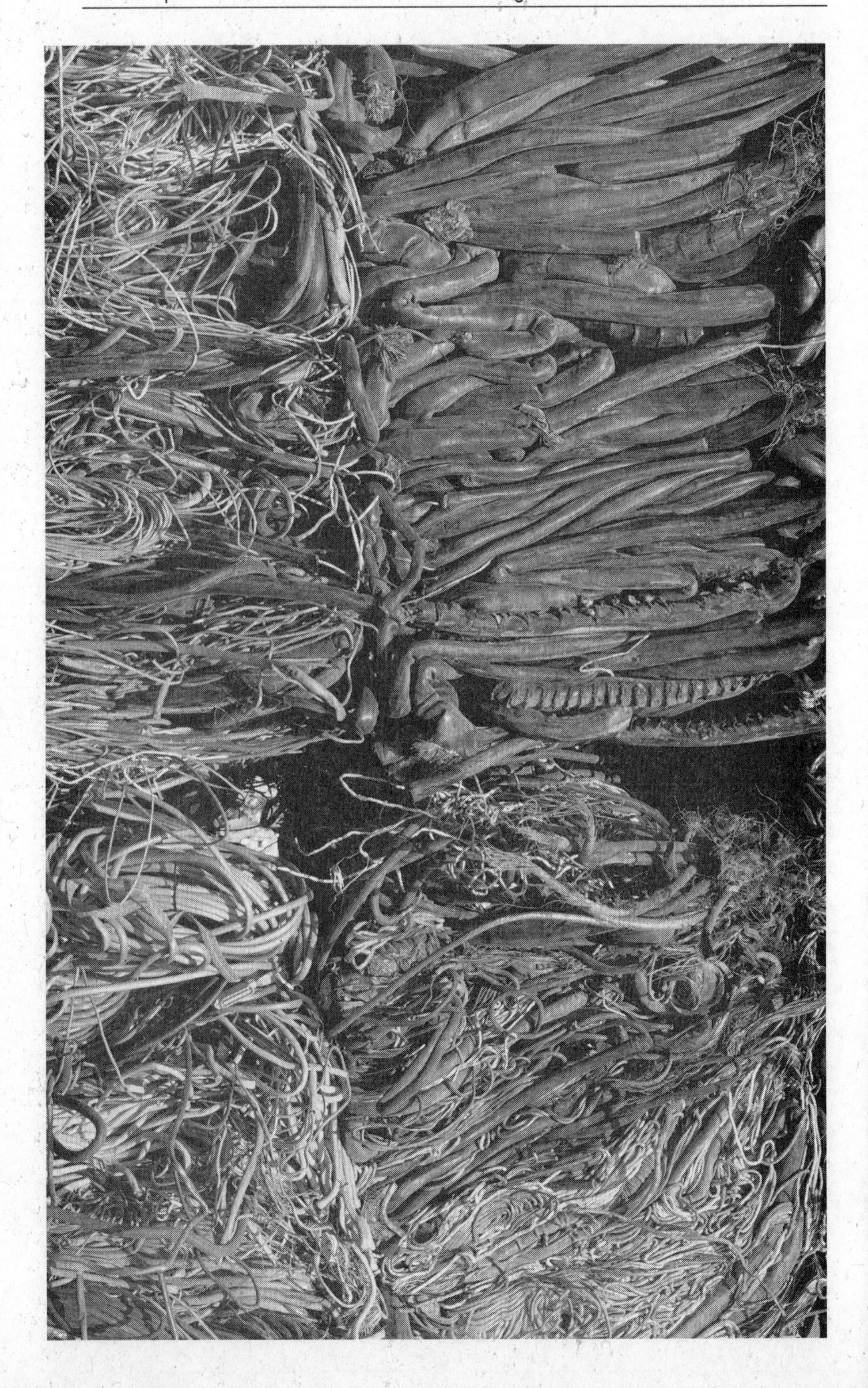

Glossary

Adobe - bricks formed from a combination of water, earth, and often straw

Batt insulation - the same fluffy, fiberglass insulation rolls most of us are familiar with, but sold in pre-cut lengths

Berm - a mound of earth or similar material often created to prevent erosion

Black water - toxic mixture of water and wastes, like urine and fecal matter

CEB - compressed earth block; an earthen building material created by a press, which offers faster production and uniformity

Clay slip - a thin mixture of clay and water with a consistency much like paint

Cob - a clump of straw, sand, and clay

Conduit - a protective casing through which cable is run

Cordwood - short (generally 1- to 2-feet long, though length can vary) round or split logs, traditionally used for firewood; a style of masonry building employing such logs

Crawlspace - an area between the foundation and the first floor of a house, typically created to house and allow access to components like plumbing

Earthbag - a bag filled with earth or similar material that is typically laid like bricks (without mortar) to form a wall

Earthship - style of house created by Michael Reynolds that uses discarded tires and passive solar design

Earth plaster - earthen material used to cover both interior and exterior walls

Emulsifier - stabilizer or strengthener added to a mixture

Footing (footer) - thick, heavy mass under the foundation that provides additional support to the structure above; typically found along perimeters and under corners

Grade - slope or incline of the land; a grade of zero would mean land that is flat; in building, "above grade" means higher than ground level, and "below grade" means underground

Gray water - non-toxic but non-drinkable water produced from sinks, showers, and dishwashers; gray water does not contain human waste like fecal matter or urine

Hygroscopic - able to take in and expel water

Infill - in alt. building, a material that goes inside a wall cavity (like insulation), or a material that bears no load placed between posts or studs

Light clay - mixture of straw and clay slip

Lintel - sturdy, often heavy beam placed above a window to bear the weight of the materials above

Load-bearing - a material, wall, or component that is carrying the weight of what is above it

Monolithic – continuous; often used in reference to a wall or foundation that is one seamless mass

Papercrete - mixture of paper (typically newspaper), water, and Portland cement

Passive solar - element that harnesses the sun's energy without effort; passive solar design is often characterized by large windows that draw in heat; components of high-thermal mass that absorb and slowly release that heat; and floor plans that allow the released heat to flow throughout the structure

Portland cement - powder produced by grinding kiln-heated raw materials (mostly limestone); the binding ingredient in traditional concretes, mortars, and stuccos

Post-and-beam - framing technique characterized by its heavy, solid members; when compared to conventional stick-framing, post-and-beam framing requires fewer, sturdier members

Rammed-earth - clay and sand mixture that is tamped or pounded to make it sturdier; traditional rammed-earth forms monolithic, sedimentary, rock-like walls; rammed-earth is also used in recycled tire construction

Screeding - rough leveling of a concrete surface

Silt - a soil component; silt particles are larger than clay particles but smaller than sand particles

Slurry - a runny to thick liquid mixture; sometimes used as a verb meaning the application of such a mixture

Soak away - a deep hole filled with gravel or similar material where non-toxic waste water (gray water) is sent to drain and be filtered

Stick-framing - type of framing used in conventional construction; characterized by its use of standardized lumber, like 2x4s or 2x6s

Subsoil - untouched soil beneath the fertile topsoil that is (hopefully) sturdy enough to build on

Tamp - to compact from above

Thermal mass - a material's ability to hold heat and release it slowly over time

Timber-framing - sometimes interchangeable with post-and-beam framing, timber framing utilizes whole or rough-sawn logs as load-bearers

Top plate - a piece of wood or similar material that sits atop a wall and creates a place for the roof to be attached; when formed of concrete, it is called a bond beam

Topsoil - the fertile, uppermost part of the soil where plants and grass grows

Troweling - leveling concrete with a trowel tool; different than screeding in that troweling produces the finished surface of the concrete

VOC - volatile organic compound; organic materials that can vaporize, enter the air, and have an effect on air quality

Bibliography

Architecture 2030 Initiative: **www.architecture2030.org**, copyright 2006-2009.

Blue Rock Station: **www.bluerockstation.com**, copyright 2003-2009.

Bynum, Richard T. and Daniel L. Rubino, *Handbook of Alternative Materials in Residential Construction.* New York: McGraw-Hill, 1999.

California Waste Management Integration Board: **http://ciwmb.ca.gov/GreenBuilding/Basics.htm#**, copyright 1995-2007.

Chiras, Daniel D. The New Ecological Home: A *Complete Guide to Green Building Options.* White River Junction, Vermont: Chelsea Green Publishing, 2004.

Community Solutions: **www.communitysolution.org**, copyright 2009.

EarthbagBuilding.com: **www.earthbagbuilding.com**, accessed 2009.

Earthship Biotecture: **www.earthship.net**, copyright 2009.

Easton, David. *The Rammed Earth House.* White River Junction, Vermont: Chelsea Green Publishing Company, 2007.

Edey, Anna. Solviva: *How to Grow $500,000 on One Acre & Peace on Earth.* Martha's Vineyard, Massachusetts: Trailblazer Press, 1998.

Elizabeth, Lynne and Cassandra Adams, ed. *Alternative Construction: Contemporary Natural Building Methods.* New York: John Wiley & Sons, 2000.

El Paso Solar Energy Association: **www.epsea.org/adobe.html**, copyright 2008.

Environmental Building News Brattleboro, VT: Building Green, June 2008.

Environmental Defense Fund: **www.edf.org**, copyright 2009.

Evans, Ianot, Michael G. Smith, and Linda Smiley. *The Hand-Sculpted House: A Practical and Philosophical Guide to Building a Cob Cottage.* White River Junction, Vermont: Chelsea Green Publishing, 2002.

Global Policy Forum: **www.globalpolicy.org/social-and-economic-policy/global-public-goods-1-101.html**, copyright 2005-2009.

Green Trust: **www.green-trust.org**, 2004-2009.

I Love Cob: **www.ilovecob.com**, copyright 2009.

LanderLand Natural Building: **www.landerland.com**, copyright 2001-2008.

Linebaugh, Peter and Marcus Rediker. *The Many-Headed Hydra: Sailors, Slaves, Commoners, and the Hidden History of the Revolutionary Atlantic.* Boston: Beacon Press, 2000.

Mehtar S, Wiid I, Todorov SD (2008). "The antimicrobial activity of copper and copper alloys against nosocomial pathogens and Mycobacterium tuberculosis isolated from healthcare facilities in the Western Cape: an in-vitro study." J. Hosp. Infect. 68 (1): 45–51. doi:10.1016/j.jhin.2007.10.009. PMID 18069086

Minke, Gernot. *Earth Construction Handbook: The Building Material Earth in Modern Architecture.* Southamptn, UK: WIT Press, 2000.

Murphy, Pat. "Post Peak – The Change Starts with Us." *Energy Bulletin.* March 22, 2006. The Community Solution. www.energybulletin.net

Natural Building Network: **www.naturalbuildingnetwork.org**, 2005-2009.

Network Earth: **www.networkearth.org**, accessed 2009.

Newcomb, Duane, *The Owner-Built Adobe House.* Albuquerque, NM: University of New Mexico Press, 2001.

New Solutions. Yellow Springs, Ohio: Community Service, Inc., Number II, January 2007.

Oikos: **www.oikos.com/library/naturalbuilding/lime.html**, copyright 2006-2009.

Passive House: **www.passivehouse.us; www.passivhaustagung.de**, copyright 2009.

Joyce, Shawn Dell. "Green Building Materials Conserve Scarce Resources." *Times Herald-Record.* November 12, 2006. **www.recordonline.com/apps/pbcs.dll/article?AID=/20061112/NEWS/611120336/-1/NEWS**

Permaculture.net: **www.permaculture.net**, accessed 2009.

Quentin Wilson: **www.quentinwilson.com/adobe-wall-construction**, copyright 2008.

Roy, Rob. *Complete Book of Cordwood Masonry House Building: The Earthwood Method.* New York: Sterling Publishing Company, Inc, 1992.

Roy, Rob. *Cordwood Building: The State of the Art.* Gabriola Island, BC: New Society Publishers, 2003.

Snell, Clarke, and Tim Callahan. *Building Green.* New York; Lark Books, 2005.

Spiegel, Ross, and Dru Meadows. *Green Building Materials: A Guide to Product Selection and Specification.* 2nd ed. New Jersey: John Wiley & Sons, 2006.

Strawbale.com: **www.strawbale.com/strawbale-faqs#pests**, copyright 2009.

The Last Straw: **www.thelaststraw.org**, 1993-2009.

Natural Builders: **www.thenaturalbuilders.com,** 2005-2009.

Toowoomba City Council Sustainable Housing Project – Costs and Benefits of Sustainable Housing Factsheet, PDF Document accessible at **www.toowoomba.qld.gov.au/index.php?option=com_docman&task=doc_view&gid=1640&Itemid=107.**

United States Department of Energy: **www.eia.doe.gov**

U.S. Private Communities Real Estate: **www.private-communities.net**, Copyright VisionDance LLC, 2003.

Weisman, Adam, and Katy Bryce. *Building with Cob: A Step-by-Step Guide.* Devon, England: Green Books, 2006.

World Watch Institute: **www.worldwatch.org/node/866**, copyright 2008.

Additional Resources

Green Builders, Designers, and Architects

- Alan Abrams, Abrams Design Build (see Case Study)
- Andrew Morrison, Straw Bale Innovations (see Case Study)
- Bill Sitkin, The ReStore (recycling), Crestone (see Case Study)
- Clarke Snell and Tim Callahan, Think Green Building: **www.thinkgreenbuilding.com**
- David Easton, Rammed Earth Works (REW Associates)
- Habib John L. Gonzalez, Sustainable Works (see Case Study)
- Jim Bryant, Vintage Copper (see Case Study)
- Suzanne Dehne, Sujo Design Inc. (see Case Study)
- Quentin Wilson, Quentin Wilson and Associates, **www.quentinwilson.com**
- Sigi Koko, Down to Earth Design (see Case Study)
- Michael Blaha, I Love Cob! (see Case Study)
- Carolyn Roberts, A House of Straw (see Case Study)
- Jai Goulding, Owner-builder (see Case Study).

- Tom Nunan, Owner-builder (see Case Study)
- Susan Nunan, Owner-builder (see Case Study)

Pictures

Michael Blaha (see Chapter 3, 9)

www.ilovecob.com

Carolyn Roberts (see Chapter 6)

www.ahouseofstraw.com

Richard Flatau (see Chapter 3)

Cordwood Construction Resources, LLC

W4837 Schulz Spur Dr

Merrill, WI 54452

www.daycreek.com/flatau

flato@aol.com

715-212-2870

Organizations

- Built It Green (regional, Bay Area): **www.builditgreen.org**
- EcoBroker: **www.ecobroker.com**
- Energy Star: **www.energystar.gov**
- Green Building Initiative: **www.thegbi.org**
- Green Map Systems (see Case Study): **www.GreenMap.org**
- LEED: **www.usgbc.org/LEED**
- Sustainable Business Alliance: **www.sustainable-building.co.nz/about.htm**
- U.S. Green Building Council (LEED requirements): **www.usgbc.org**

Conventions

Greenbuild: **www.greenbuildexpo.org**

Green Festivals: **www.greenfestivals.org**

Periodicals

The Last Straw Journal

Mother Earth News

Natural Home Magazine

Green Building News

Environmental Building News

Active Rain

The Green Guide

Web sites

Construction guide: **www.toolbase.org**

Earthbag and papercrete: **www.greenhomebuilding.com/QandA/earthbag/fillingandlaying.htm**

Earth-based building materials such as cob, pallets, straw: **www.worldhandsproject.org**

Fair Trade Federation (FTF) for locating products: **www.fairtradefederation.org**

Green building discussion lists: **http://repp.org/discussion-groups/index.html**

Plaster pigments (natural): **www.sinopia.com**

Paints and sealers: **www.afmsafecoat.com**

Rammed earth, plus many links: **www.diyrammedearth.com**

Recycling materials for building: **www.greenhomebuilding.com/recyclematerials.htm**

Recyclable materials at C&D sites: **www.recycleworks.org/con_dem/recycling_faq.html**

Supplies: **www.greenbuildingsupply.com**

Author Biography

Jon Nunan

Jon Nunan is a writer for the home improvement industry, and staunch supporter of learning by doing and letting that which does not matter truly slide. When not writing, he can be found identifying wild mushrooms, picking berries, and taking on small projects in and around the wood pile he currently calls home.

Index

A

B

C

D

E

Equipment, 101, 103, 173, 217-218, 21

F

G

H

I

L

M

N

P

R

Rafters, 256, 111-112, 114, 119, 126, 241

S

T

W